Cybersecurity Lexicon

Luis Ayala

Apress®

Cybersecurity Lexicon

ISBN-13 (pbk): 978-1-4842-2067-2

ISBN-13 (electronic): 978-1-4842-2068-9

Managing Director: Welmoed Spahr
Acquisitions Editor: Susan McDermott
Developmental Editor: Douglas Pundick
Technical Reviewer: Tito Cordera
Editorial Board: Steve Anglin, Pramila Balen, Louise Corrigan, James DeWolf, Jonathan Gennick, Robert Hutchinson, Celestin Suresh John, Nikhil Karkal, James Markham, Susan McDermott, Matthew Moodie, Douglas Pundick, Ben Renow-Clarke, Gwenan Spearing
Coordinating Editor: Rita Fernando
Copy Editor: Kim Burton-Weisman
Compositor: SPi Global
Indexer: SPi Global

Distributed to the book trade worldwide by Springer Science+Business Media New York, 233 Spring Street, 6th Floor, New York, NY 10013. Phone 1-800-SPRINGER, fax (201) 348-4505, e-mail orders-ny@springer-sbm.com, or visit www.springer.com. Apress Media, LLC is a California LLC and the sole member (owner) is Springer Science + Business Media Finance Inc (SSBM Finance Inc). SSBM Finance Inc is a Delaware corporation.

For information on translations, please e-mail rights@apress.com, or visit www.apress.com.

Apress and friends of ED books may be purchased in bulk for academic, corporate, or promotional use. eBook versions and licenses are also available for most titles. For more information, reference our Special Bulk Sales–eBook Licensing web page at www.apress.com/bulk-sales.

Any source code or other supplementary materials referenced by the author in this text is available to readers at www.apress.com. For detailed information about how to locate your book's source code, go to www.apress.com/source-code/.

I want to thank my wife, Paula, who has been with me through thick and thin for the last 35 years. I also want to thank our son, Christopher.

Contents

About the Author

Luis Ayala worked for the US Department of Defense for more than 25 years, with the past 11 years at the Defense Intelligence Agency. Prior to his appointment as a defense intelligence senior leader in 2008, he held several leadership positions at the branch and division levels.

His tenure culminated with the position as senior technical expert (facilities/construction). Mr. Ayala earned his Bachelor of Architecture degree from Pratt Institute and he received his Master of Science and Technology Intelligence from the National Intelligence University. NIU is the intelligence community's sole accredited, federal degree granting institution. His master's thesis, titled "Cybersecure Facilities for the Intelligence Community," is classified. Mr. Ayala was awarded the DIA Civilian Expeditionary Medal and the Civilian Combat Support Medal.

About the Technical Reviewer

Tito Cordero Jr. has 40 years of experience in the US Department of Defense and the private sector. Prior to his retirement, Mr. Cordero was a special agent with the Defense Security Service (DSS). His main responsibility was computer security within the industrial contractor sector. During his stay at DSS, Mr. Cordero provided expert advice on computer security and conducted espionage investigation for the Department of Defense.

Mr. Cordero worked for Defense Intelligence Agency, providing computer support and managing the implementation of the first US Sensitive Compartmental Facility (SCI) installation outside the United States, in Iraq. During this time, Mr. Cordero served as the lead computer expert to direct the installation of computers, networks, and security protocols for this effort.

Mr. Cordero earned his Bachelor of Business Management with a minor in computer science from the University of Phoenix at Arizona. He as earned a number of awards from DoD, including the Army's Superior Civilian Service Award from Office of Secretary of Defense. He was awarded the Joint Civilian Service Commendation Award for services rendered during the Iraq War in 2003, which was a direct effort in his part for the implementation of the SCI for the Defense Intelligence Agency. Mr. Cordero also was awarded the Desert Shield/Desert Storm medal for contribution to the War in 1990 as a direct support systems engineer while working for Army Signal Command as a staff officer.

Preface

Hackers are estimated to control more than 10% of the computers on the Internet.

Discussion of building hacks and cyber-attacks is clouded by a lack of standard definitions and a general misunderstanding about how bad actors can actually employ cyber technology as a weapon in the real world. The Cybersecurity Lexicon was developed specifically to introduce building managers to the vulnerability of industrial control systems (and SCADA systems) to cyber-physical attack. Architects, engineers, and facility engineers need to know how to defend their buildings against cyber-physical attack by learning more about the cyber "attack surface," which is the sum of all the "attack vectors."

This desk reference offers easy-to-understand definitions of cyber jargon and technical terms related to automated control systems common to buildings, utilities, and industry. This book is not intended for cyber-professionals that (hopefully) already know most of this information. Although written primarily to focus on building controls and cyber-physical attacks, much of the terminology applies to cyber-attacks in general. The book is a handy desk reference for architects, engineers, building managers, students, researchers, and consultants interested in preventing cyber-physical attacks against their facilities in the real world.

CHAPTER 1

# #

2-pipe HVAC (heating, ventilating, air conditioning) system: A two-pipe system consists of fan coil units with single coils connected to two pipes. The two pipes, one supply and one return, are connected to supply lines in the building's mechanical room. Supply lines can supply hot water or chilled water. Two pipe systems are less flexible than a four-pipe system because the entire building is in either heating mode or cooling mode.

3-level password protection: Requiring an additional password to authorize critical operations greatly reduces the surface area to attack the secondary credentials since they are used less often; for example, a power-on password, a parameter-setting password, and a parameter correction password. Requiring reauthentication to perform special actions can protect against CSRF (cross-site request forgery) attacks. Cross-site request forgery is also known as *one-click attack* or *session riding*.

3D laser scanning system: A very precise system that captures 3D shapes on an assembly line to inspect, measure, and collect data of real-world objects.

4 -pipe HVAC system: A four-pipe system includes the distribution system that consists of a hot water supply with return lines and a chilled water supply with return lines. Four-pipe systems can supply heat to one room while simultaneously cooling another room. A four-pipe HVAC system can be hacked so the building heat and cooling systems are on at the same time, working against each other—driving up energy costs.

© Luis Ayala 2016
L. Ayala, *Cybersecurity Lexicon*, DOI 10.1007/978-1-4842-2068-9_1

CHAPTER 2

A

ABC fire extinguisher: Chemically-based fire extinguishing device used to put out ordinary combustible, flammable liquid and electrical fires.

Abnormal Situation Management (ASM): ASM was developed with the goal of improving safety and performance in process plants. Over the past 20 years, the ASM Consortium has developed knowledge, tools, and products designed to prevent and manage abnormal situations in the process industry. This knowledge is directly applicable to cyber-physical attack response.

abort gate: A high-speed damper designed to divert sparks, flames, smoke, combustion gases, and burning material out of a pneumatic system and evacuate the air through a discharge hood. An abort gate contains a spring-assisted blade that is usually held in place by an electromagnet that typically will react within 1/2 second. Used to protect dust collection systems and prevent dust explosions. If an attacker hacks a dust collection system's abort gate at an industrial facility to prevent it from doing its job, there is a high probability of a dust explosion.

absolute encoder: Maintains equipment position information when power is lost from a manufacturing environment. Once power is restored, the position information is immediately available.

acceptable level of risk: Typically refers to the point at which the level of risk is more acceptable than the cost to mitigate the risk (in dollars or effect on building mission function).

access: (1) The technical ability to do something with a computer resource. This usually refers to ability to read, create, modify, or delete a file; execute a program; or use an external connection, admission, or entrance. (2) The ability or opportunity to obtain sensitive or classified information. SOURCE: CNSSI-4009

access control: The limiting of access to the resources of an IT (information technology) system only to authorized users, programs, processes, or other IT systems. SOURCE: FIPS 201; CNSSI-4009

access control list (ACL): A list of permissions that specifies which users or system processes are granted access to objects. Each entry specifies a subject and an operation (Joe: write, read, Sam: read only). SOURCE: CNSSI-4009

Access Matrix: An Access Matrix uses rows to represent subjects and columns to represent objects with privileges listed in each cell.

access point: A device that logically connects wireless client devices operating in infrastructure to one another and provides access to a distribution system, if connected, which is typically an organization's enterprise wired network. SOURCE: SP 800-48; SP 800-121

access profile: The association of a user with a list of protected objects that the user may access. SOURCE: CNSSI-4009

access type: The privilege to perform action on an object. Read, write, execute, append, modify, delete, and create are examples of access types. SOURCE: CNSSI-4009

account harvesting attack: The process of collecting all the user account names on a computer network. Often used to refer to computer spammers, individuals who try to sell or seduce others through e-mail advertising or solicitation. Account harvesting involves using computer programs to search areas on the Internet in order to gather lists of e-mail addresses from a number of sources, including chat rooms, domain names, instant message users, message boards, newsgroups, online directories for web pages, web pages, and other online destinations.

account management: Involves the process of requesting, establishing, issuing, and closing user accounts; tracking users and their respective access authorizations; and managing these functions. SOURCE: SP 800-12

accountability: The property that enables activities on a system to be traced to individuals, who may then be held responsible for their actions. The security goal that generates the requirement for actions of an entity to be traced uniquely to that entity. This supports non-repudiation, deterrence, fault isolation, intrusion detection and prevention, and after-action recovery and legal action. SOURCE: SP 800-27

ACK: In some digital communications protocols, notification that a signal has been received successfully. The ACK signal is sent by the receiving station after receipt of data. When the source gets an ACK signal, it transmits the next block of data.

ACK piggybacking attack: Is an active form of wiretapping when a hacker sends an ACK inside another packet to the same destination.

active cyber-attack: An intentional cyber-attack perpetrated that attempts to alter a SCADA (supervisory control and data acquisition) system, its resources, its data, or its operations. SOURCE: CNSSI-4009

active attack: An attack on the authentication protocol where the attacker transmits data to the claimant, credential service provider, verifier, or relying party. Examples of active attacks include man-in-the-middle, impersonation, and session hijacking. SOURCE: SP 800-63

active content: Electronic documents that can carry out or trigger actions automatically on a computer platform without the intervention of a user. Software in various forms that is able to automatically carry out or trigger actions on a computer platform without the intervention of a user. SOURCE: SP 800-28, CNSSI-4009

activation: When all or a portion of the cyber-physical attack recovery plan has been put into motion.

active security testing: Security testing that involves direct interaction with a target, such as sending packets to a target. SOURCE: SP 800-115

activities: An assessment object that includes specific protection-related pursuits or actions supporting an information system that involve people (e.g., conducting system backup operations, monitoring network traffic). SOURCE: SP 800-53A

activity monitor: Used to prevent a cyber-attack by monitoring a system and blocking malicious activity. A system monitor for an operating system, which also incorporates task manager functionality. Some of the functions include killing computer processes, viewing CPU load, checking the amount of random access memory in use, and other functions.

actuator: A pneumatic, hydraulic, or electrically powered device that supplies force and motion so as to position a valve's closure member at or between the open or closed position. Used on electronic damper actuators, valve actuators, relays, and so forth.

ad hoc network: A wireless network that dynamically connects wireless client devices to each other without the use of an infrastructure device, such as an access point or a base station. SOURCE: SP 800-121

add-on security: Incorporation of new hardware, software, or firmware safeguards in an operational information system. SOURCE: CNSSI-4009

adequate security: Security commensurate with the risk and magnitude of the harm resulting from the loss, misuse, or unauthorized access to, or modification of information. This includes assuring that systems and applications used by owner operate effectively and provide appropriate confidentiality, integrity, and availability through the use of cost-effective management, personnel, operational, and technical controls. SOURCE: SP 800-53; FIPS 200; OMB Circular A-130, App. III

address space probe attack: When a hacker is looking for security holes that might be exploited he first will attempt to map IP address space.

addressing: Ceiling light fixture ballast and drivers are centrally addressable through the lighting central control software. In emergency mode, all light fixtures immediately adjust lights to full light output and remain in that mode until that mode is deactivated.

administrative safeguards: Administrative actions, policies, and procedures to manage the selection, development, implementation, and maintenance of security measures to protect electronic health information and to manage the conduct of the covered entity's workforce in relation to protecting that information. SOURCE: SP 800-66

Advanced Encryption Standard (AES): An encryption standard developed by the National Institute of Standards and Technology (NIST) in 2001. SOURCE: FIPS 197; CNSSI-4009

Advanced Metering Infrastructure (AMI): Refers to the full measurement and collection system that includes meters at the customer site, communication networks between the customer and a service provider, such as an electric, gas, or water utility, and data reception and management systems that make the information available to the service provider. Systems are comprised of state-of-the-art electronic/digital hardware and software that combines interval data measurement with continuously available remote communications. These systems enable measurement of detailed, time-based information and frequent collection and transmittal of such information. SOURCE: Electric Power Research Institute

Advanced Persistent Threats (APT): An adversary that possesses sophisticated levels of expertise and significant resources that allow it to create opportunities to achieve its objectives by using multiple attack vectors (e.g., cyber, physical, and deception). These objectives typically include establishing and extending footholds within the information technology infrastructure of the targeted organizations for purposes of exfiltrating information, undermining or impeding critical aspects of a mission, program, or organization; or positioning itself to carry out these objectives in the future. The advanced persistent threat (1) pursues its objectives repeatedly over an extended period of time; (2) adapts to defenders' efforts to resist it; and (3) is determined to maintain the level of interaction needed to execute its objectives. An unauthorized person gains undetected access to a system and stays for a long period of time. The intent is to steal data. A persistent presence is sometimes called *consolidation*. APTs can wait a long time before becoming active. By performing a gap analysis of the network configuration, hidden APTs can be made to show themselves either by detection methods or making them become visible by exposing themselves through their designed behavior. SOURCE: SP 800-39

advisory: Notification of significant new trends or developments regarding the threat to the information systems of an organization. This notification may include analytical insights into trends, intentions, technologies, or tactics of an adversary targeting information systems. SOURCE: CNSSI-4009

adware: Is a form of software that has advertisements embedded in the application. Adware is similar to malware, because it has been known in the past to infect computers with viruses.

airbrick: A special clay brick with perforations designed to provide high levels of ventilation into a building's exterior cavity walls and underfloor voids.

air conditioning systems: Most buildings have one of two types of central air conditioning: *direct expansion* (DX) type and *chilled water* type. A DX system passes the air used for cooling the room directly over the cooling coil. In a chilled water system, a refrigeration system first chills the water, which is then used to chill the air.

Aircrack-Ng: A set of tools for auditing wireless networks. Aircrack-ng is an 802.11 WEP and WPA-PSK keys cracking program that can recover keys once enough data packets have been captured. Aircrack-ng implements the standard FMS attack, making the attack much faster compared to other WEP cracking tools.

Airdrop-Ng: A program used for targeted, rule-based deauthentication of users. It can target based on MAC address, type of hardware, or completely deauthenticate ALL users by the transmission of deauthentication packets.

airflow measuring station: Consists of single or multiple airflow elements, factory-mounted and pre-piped in the HVAC air duct. Airflow sensing elements report actual airflow in feet per minute.

air gap: One or more computers are physically isolated from unsecured computer networks, such as the public Internet or an unsecured local area network for security purposes. Air gap computers are not connected by wire or wirelessly, and (generally) cannot communicate directly with each other.

air handling unit (AHU): A device used to regulate and circulate air as part of a HVAC system. An air handler is a large metal container having a blower, heating or cooling elements, filter racks or chambers, sound attenuators, dampers and UV light to kill bacteria. Air handlers usually connect to a ductwork ventilation system that distributes conditioned air throughout the building and returns it to the AHU. Sometimes AHUs discharge (*supply*) and admit (*return*) air directly to the space without ductwork. A large air handler that conditions 100% outside air, and no recirculated air, is a makeup air unit (MAU). An air handler designed for outdoor use, typically on roof, is a package unit (PU) or rooftop unit (RTU). Small air handlers, called *terminal units*, may only include an air filter, coil, and blower; simple terminal units are called *blower coils* or *fan coil units*.

air traffic control (ATC): Ground-based controllers who direct aircraft on the ground and through controlled airspace to prevent collisions, organize and expedite the flow of traffic, and provide advisory service to aircraft in non-controlled airspace. In some countries, they are operated by the military.

alarm filtering: Sometimes failure of one piece of building equipment may cause another piece of equipment to fail. Alarm filtering reports the original failure with more priority than subsequent failures showing the technician which equipment to repair first.

alarm flooding: The annunciation of more alarms in a given period of time than a human operator can respond. Ten alarms per minute are typically the most that a technician can handle.

alarm indication station (AIS): Building control system device with an indication light, alarm horn, and an alarm horn silence switch. Upon an alarm condition, the problem indicator will light and the horn sounds. The horn can be silenced through the alarm silence switch, while the visual indication will continue until the condition has been corrected. A power interruption or open alarm contact typically resets the AIS.

alarm data: At a minimum, the following is displayed and stored for a BCS:

- Identification of building under alarm including, building control system (or subsystem) and device.

- Date and time of occurrence.

- Alarm type:

  a. **Unreliable**: The device has failed due to bad data or the sensing device or alarm parameter was out-of-range.

  b. **High-level alarm**.

  c. **Low-level alarm**.

- Current status of the alarm point.

- Alarm limits.

- Alarm priority.

- Alarm Message: Unique message. Assignment of messages to an alarm can be edited when setup.

- Acknowledgement status including the time, date and user name.

alarms: Building monitoring and control software is capable of generating alarms by comparing the value of any point to user-configurable limits. There are two alarm priority levels:

- **Critical alarms** remain in alarm until the alarm condition no longer exists.

- **Informational alarms** remain in alarm until the alarm is acknowledged.

alarm replay: A fire alarm control feature that allows emergency response personnel to determine the sequence in a multiple zone fire to determine the origin and progress of the fire.

alarm storage and reports: Building control software stores each alarm and alarm data to a hard drive and retains the information. Stored data is sortable and searchable.

alert: Notice that a cyber-attack has occurred requiring activation of a recovery plan. SOURCE: CNSSI-4009

algorithm: A set of step-by-step instructions for a computer procedure.

allocation: The process an organization employs to determine whether security controls are defined as system-specific, hybrid, or common. The process an organization employs to assign security controls to specific information system components responsible for providing a particular security capability (e.g., router, server, remote sensor). SOURCE: SP 800-37

all-source intelligence: Analyzing the threat information from multiple sources, disciplines, and agencies across the Intelligence Community. Synthesizing intelligence information in context and drawing insights about possible implications.

alternate work site: A location (other than the usual facility) used to process data and conduct business in the event of a cyber-attack. SOURCE: CNSSI-4009

amplification attack: A reflected DDoS attack when a single UDP packet generates tens or hundreds of times the bandwidth to overwhelm a building control system with DNS response traffic. A denial-of-service technique that uses numerous hosts. SOURCE: SP 800-61; CNSSI-4009

anomaly-based detection: The process of comparing definitions of what activity is considered normal against observed events to identify significant deviations. SOURCE: SP 800-94

anonymizer: An anonymous proxy that attempts to make activity on the Internet untraceable. It is a proxy server computer that acts as an intermediary and privacy shield between a client computer and the rest of the Internet. It accesses the Internet on the user's behalf, protecting personal information by hiding the client computer's identifying information. Proxies can be daisy chained. Chaining anonymous proxies can make traffic analysis far more complex and costly by requiring the eavesdropper to be able to monitor different parts of the Internet. An anonymizing re-mailer can use this concept by relaying a message to another re-mailer, and eventually to its destination.

anti-jam: Countermeasures ensuring that transmitted information can be received despite deliberate jamming attempts. SOURCE: CNSSI-4009

anti-router: A device that detects Wi-Fi surveillance devices and blocks them from accessing your Wi-Fi network. Every wireless device has a unique hardware signature assigned to it by the manufacturer. These signatures are broadcast by wireless devices as they probe for, connect to, and use wireless networks. An anti-router "sniffs" the airwaves for these signatures, looking for surveillance devices such as a drone. If a banned device is discovered an alarm is triggered and if that device is connected to a network that the anti-router is trained to defend, a stream of "de-authentication packets" are sent automatically to disconnect the rogue device.

anti-spoof: Countermeasures taken to prevent the unauthorized use of legitimate *identification and authentication* (I&A) data, however it was obtained, to mimic a subject different from the attacker. SOURCE: CNSSI-4009

antispyware software: A program that specializes in detecting both malware and non-malware forms of spyware. SOURCE: SP 800-69

antivirus software: A program that monitors a computer or network to identify all major types of malware and prevent or contain malware incidents. SOURCE: SP 800-83

applet attack: A Java general-purpose computer programming language that uses the client's web browser to provide a user interface and disable the Java security sandbox.

appliance hacks: With the growth of the Internet of Things, even common appliances such as dishwashers, coffee makers, clothes dryers, and baby monitors connect to the Internet and could be used to gather intelligence. Manufacturers use this capability to troubleshoot performance of their equipment, monitor usage, and improve the customer "experience." Unfortunately, knowing when you use the appliance provides data could help hackers geolocate your current position, and learn your habits and schedule.

application: The use of computer program designed to assist people perform a specific set of requirements.

Application Generic Controller (AGC): A device furnished with a (limited) preloaded application that can also be reprogrammed. ProgramID and XIF file of the AGC are fixed.

application program: A software program hosted by an information system. Software program that performs a specific function directly for a user and can be executed without access to system control, monitoring, or administrative privileges. SOURCE: SP 800-37; CNSSI-4009

Application Specific Controller (ASC): An intelligent electronic device with a built-in application that is configurable, but not reprogrammable.

application recovery: The restoration of building control system software and data, after a computer has been restored or replaced.

application whitelist: A list or register of entities that are being provided a particular privilege, service, mobility, access, or recognition. Entities on the list will be accepted, approved, and/or recognized. Whitelisting is the reverse of blacklisting, the practice of identifying entities that are denied, unrecognized, or ostracized.

Application Whitelisting (AWL): AWL can detect and prevent attempted execution of malware uploaded by adversaries. The static nature of BCS and SCADA systems make them ideal candidates for AWL.

approved security function: A security function (e.g., cryptographic algorithm, cryptographic key management technique, or authentication technique) that is either: specified in an Approved Standard; or adopted in an Approved Standard and specified either in an appendix of the Approved Standard or in a document referenced by the Approved Standard; or specified in the list of Approved security functions. SOURCE: FIPS 140-2

arc fault: A high-power electricity discharge between two conductors. The heat can break down the wire's insulation and possibly trigger an electrical fire. An arc fault can range from a few amps up to thousands of amps and are highly variable in strength and duration. Common causes of arc faults include faulty connections due to corrosion or faulty installation. Arc fault circuit breakers quickly detect increases in current draw.

arc fault circuit interrupter (AFCI): A duplex receptacle or circuit breaker that breaks the circuit to prevent electrical fires when it detects a dangerous electrical arc. An AFCI distinguishes between a harmless arc that occurs during normal operation and an undesirable arc that can occur due to a broken conductor.

area lighting controller (ALC): Provides high-power switching and dimming interface between a group of lights and the BCS network.

ARP spoofing attack: A technique by which an attacker sends (spoofed) Address Resolution Protocol (ARP) messages onto a local area network. Generally, the aim is to associate the attacker's MAC address with the IP address of another host, such as the default gateway, causing any traffic meant for that IP address to be sent to the attacker instead. ARP spoofing may allow an attacker to intercept data frames on a network, modify the traffic, or stop all traffic. Often the attack is used as an opening for other attacks, such as denial-of-service, man in the middle, or session hijacking attacks. Also called *ARP cache poisoning* or *ARP poison routing*.

ARPANET: Advanced Research Projects Agency Network, an early packet-switched network built under contract to the Advanced Research Project Agency (ARPA) a United States Department of Defense agency that led to the development of the Transmission Control Protocol / Internet Protocol or Internet.

array: An arrangement of two or more disk drives that are configured in a Redundant Array of Inexpensive Disks (RAID) or daisy-chain fashion.

arrival manager (AMAN): A system aid for the ATC at airports that calculates planned arrivals to maintain optimal throughput at the runway, reduce arrival queuing and distribute arrival information.

artifact: The digital remnants of a cyber-attack or incident activity. These could be software that was used by a hacker, a collection of tools, malicious code, logs, files, output from tools, or status of a building control system after a cyber-attack. Examples range from Trojan-horse programs and computer viruses to programs that exploit vulnerabilities or objects of unknown type and purpose found on a compromised computer.

assessment findings: Assessment results produced by the application of an assessment procedure to a security control or control enhancement to achieve an assessment objective; the execution of a determination statement within an assessment procedure by an assessor that results in either a satisfied or other than satisfied condition. SOURCE: SP 800-53A

asset: (1) A value placed on goods. (2) A major application, general support system, high impact program, physical plant, mission critical system, personnel, equipment, or a logically related group of systems. SOURCE: CNSSI-4009

assumptions: Something presumed to be the case about unknown disaster situations that a cyber-physical attack recovery plan is based on.

assurance: Grounds for confidence that the other four security goals (integrity, availability, confidentiality, and accountability) have been adequately met by a specific implementation. "Adequately met" includes (1) functionality that performs correctly, (2) sufficient protection against unintentional errors (by users or software), and (3) sufficient resistance to intentional penetration or bypass. SOURCE: SP 800-27

assurance case: A structured set of arguments and a body of evidence showing that an information system satisfies specific claims with respect to a given quality attribute. SOURCE: SP 800-53A; SP 800-39

assured information sharing: The ability to confidently share information with those who need it, when and where they need it, as determined by operational need and an acceptable level of security risk. SOURCE: CNSSI-4009

assured software: Computer application that has been designed, developed, analyzed, and tested using processes, tools, and techniques that establish a level of confidence in it. SOURCE: CNSSI-4009

asymmetric keys: Two related keys, a public key and a private key that are used to perform complementary operations, such as encryption and decryption or signature generation and signature verification. SOURCE: FIPS 201

Asynchronous Transfer Mode (ATM): A network architecture that establishes a connection between the originating and receiving stations allowing transmission of data (video, audio, etc.) over one line. ATM was developed to meet the needs of the Broadband Integrated Service Digital Network.

attack: An attempt to gain unauthorized access to building control system services, resources, or information, or an attempt to compromise building control system integrity and availability. SOURCE: SP 800-32

attack pattern: Similar cyber events or behaviors that may indicate that a cyber-attack is occurring or has occurred.

attack sensing and warning (AS&W): Detection, correlation, identification, and characterization of intentional unauthorized activity with notification to decision makers so that an appropriate response can be developed. SOURCE: CNSSI-4009

attack signature: A specific sequence of events indicative of an unauthorized access attempt. A characteristic byte pattern used in malicious code or an indicator or set of indicators that allows the identification of malicious network activities. SOURCE: CNSSI-4009; SP 800-12

attack surface: The sum of all the *attack vectors*, where a hacker can attempt to enter or extract data from a building control system.

attack tools: Hackers use attack tools that leverage Google, Bing, and other search engines to find information and expose vulnerabilities of building control systems.

attack tree: A conceptual diagram showing how a computer system might be attacked by describing the threats and possible cyber-attacks to realize those threats. Cyber-attack trees lend themselves to defining an information assurance strategy and are increasingly being applied to industrial control systems and the electric power grid. Executing a strategy changes the attack tree.

attack vectors: This is a path or means by which a hacker or cracker can gain access to a computer or network server in order to deliver a payload or malicious outcome. Ways in which your BCS or CMMS can be attacked:

- **Internet access**: If your BCS is connected, your network has already been scanned and mapped.

- **wireless network**: If you use wireless devices on your BCS, it has already been scanned and mapped.

- **insider threat**: Deliberate or inadvertent activity.

- **direct-access attack**: Gaining physical access to a BCS network device.

- **removable media**: USB, floppy, CD, laptop, anything that can connect directly to a BCS network device.

- **e-mail**: Malware delivered by phishing e-mail such as a virus, Trojan horse, worm.

- **other networks**: A connection to the enterprise network can be one way to get into the BCS.

- **supply chain**: If it's made overseas, it's probably got some hidden program you'll never find.

- **improper installation or usage**: Deliberate or inadvertent activity.

- **theft of equipment**: Lose a vital piece of equipment and your system can be left defenseless.

- **cyber-drone**: A drone can monitor a facility seeking wireless signals such as from network printers.

- **other**: Whatever I left out.

attorney-induced symptomology: Symptoms that arise when a possible lawsuit may be involved.

attribute authority: An entity, recognized by the Federal Public Key Infrastructure (PKI) Policy Authority or comparable agency body as having the authority to verify the association of attributes to an identity. SOURCE: SP 800-32

attribute-based access control: Access control based on attributes associated with and about subjects, objects, targets, initiators, resources, or the environment. An access control rule set defines the combination of attributes under which an access may take place. SOURCE: SP 800-53; CNSSI-4009

attribute-based authorization: A structured process that determines when a user is authorized to access information, systems, or services based on attributes of the user and of the information, system, or service. SOURCE: CNSSI-4009

ATFP emergency shutdown switch: Anti-Terrorism Force Protection feature used to quickly turn off a building HVAC system upon detection of possible chemical, biological or radiological threat.

audit: Independent review and examination of records and activities to assess the adequacy of system controls, to ensure compliance with established policies and operational procedures, and to recommend necessary changes in controls, policies, or procedures. SOURCE: SP 800-32 Independent review and examination of records and activities to assess the adequacy of system controls, to ensure compliance with established policies and operational procedures. SOURCE: CNSSI-4009

audit data: Chronological record of system activities to enable the reconstruction and examination of the sequence of events and changes in an event. SOURCE: SP 800-32

audit log: A chronological record of system activities. Includes records of system accesses and operations performed in a given period. SOURCE: CNSSI-4009

audit reduction tools: Preprocessors designed to reduce the volume of audit records to facilitate manual review. Before a security review, these tools can remove many audit records known to have little security significance. These tools generally remove records generated by specified classes of events, such as records generated by nightly backups. SOURCE: SP 800-12; CNSSI-4009

audit review: The assessment of an information system to evaluate the adequacy of implemented security controls, assure that they are functioning properly, identify vulnerabilities, and assist in implementation of new security controls where required. This assessment is conducted annually or whenever significant change has occurred and may lead to recertification of the information system. SOURCE: CNSSI-4009

audit system: An independent review, examination of records and activities to access the adequacy of building controls to ensure compliance with policies and procedures. SOURCE: SP 800-32

audit trail: A series of records of computer events about a building controls system, operating system, an application, or user activities during a given period. SOURCE: SP 800-47

auditing: The information gathering, review, and analysis of management, operations, and controls to ensure policy compliance and security from vulnerabilities.

Aurora vulnerability hack: In 2007, the Idaho National Laboratory (INL) conducted a test to demonstrate how a cyber-physical attack could destroy physical components of the electric grid. INL used a computer program to rapidly open and close a diesel generator's circuit breakers out of phase from the rest of the grid. Every time the breakers were closed, the torque from the synchronization caused the generator to bounce and shake, eventually causing parts of the generator to be ripped apart and sent flying as far as 80 feet. This vulnerability can be mitigated by preventing out-of-phase opening and closing of the circuit breakers. A cyber-physical attack that takes down the commercial power grid will cause a rise in mortality rates as health and safety systems fail, a drop in trade as ports shut down, and disruption to transport and infrastructure.

authenticate: To confirm the identity of an entity when that identity is presented. SOURCE: SP 800-32 To verify the identity of a user, user device, or other entity. SOURCE: CNSSI-4009

authentication: Verifying the identity of a user, process, or device, often as a prerequisite to allowing access to resources in an information system. SOURCE: SP 800-53; SP 800-53A; SP 800-27; FIPS 200

authentication mechanism: Hardware-or software-based mechanisms that force users to prove their identity before accessing data on a device. SOURCE: SP 800-72; SP 800-124

authentication period: The maximum acceptable period between any initial authentication process and subsequent reauthentication processes during a single terminal session or during the period data is being accessed. SOURCE: CNSSI-4009

authentication protocol: A defined sequence of messages between a Claimant and a Verifier that demonstrates that the Claimant has possession and control of a valid token to establish his/her identity, and optionally, demonstrates to the Claimant that he or she is communicating with the intended Verifier. SOURCE: SP 800-63

authentication token: Authentication information conveyed during an authentication exchange. SOURCE: FIPS 196

authorization: The official management decision given by a senior organizational official to authorize operation of an information system and to explicitly accept the risk to organizational operations (including mission, functions, image, or reputation), organizational assets, individuals, other organizations, and the Nation based on the implementation of an agreed-upon set of security controls. SOURCE: SP 800-53; SP 800-53A; CNSSI-4009; SP 800-37

auto answer: A feature that enables a fax machine or modem to automatically answer a telephone call.

auto-hacking attack: An easy-to-use device with the auto-hacking function will hack into a Wi-Fi network without a computer. Simply turn on the device, select a network and the device will hack it automatically. It is a standalone machine and does not require boot from disc or computer.

auto recloser: A recloser, or autorecloser, is a circuit breaker that can automatically close after it has been opened due to an electrical fault. They are used on overhead electrical distribution systems to detect and interrupt momentary faults. Many short-circuits on overhead cables clear themselves, a recloser automatically restores power to the line after a momentary fault. Outages may cause building controls to lose time settings, lose data in volatile memory, halt, restart, or be damaged.

automated: Synonym: *computerized*. Using machinery or electronics to perform most tasks.

automated password generator: An algorithm that creates random passwords that have no association with a particular user. SOURCE: FIPS 181

automatic control valves: Flow control valves respond to signals generated by intelligent electronic devices such as flow meters or temperature gauges and automatically regulate the flow or pressure of fluids. A hacker can cut or slow fluid flow using the BCS.

autonomous building control system: A computer network or series of networks that are all under one administrative control. Sometimes called a routing domain. An autonomous system is assigned a globally unique number referred to as an *Autonomous System Number (ASN)*.

autonomous weapons systems (AWS): "Killer robots." A machine operated by a computer utilizing artificial intelligence software. The machine consists of a variety of sensors, motors, batteries, memory devices, computational devices, and wiring. The one thing these items have in common is they all rely on electricity to communicate and control the actions of the AWS. Because of the complexity required for an AWS, the machine's devices operate like nodes on a computer network with each component having an "address." Commands generated by the AI are communicated through the network to the device needed to perform a particular function. So, for example, if the AWS decides to run south to attack a target, the actions needed to "run" are preprogrammed so the AWS travels as directed. Variables such as direction, speed, and obstacle avoidance are all choreographed carefully to obtain the desired result. Once the AWS reaches the destination, the AI seeks valid targets and selects the proper weapon for the lethality level dictated by the situation. Resisting forces are dispatched quickly and surrendering soldiers are "de-weaponized" and "captured."

Assuming the AWS relies on a command and control system to receive orders and report status, the communications medium is one possible entry pathway for cyber-attack. For example, a man-in-the-middle cyber-attack might intercept a valid command, analyze the command, and send a different command to the AWS. So, for example if the host wants the AWS to run south, a spoofed command could send the AWS running north. Other spoofed messages could tell the AWS to damage itself (jump off a cliff), empty its weapons, shoot each other or simply shut down wherever it happens to be. It may be possible to use "social engineering" against an AWS (fool the artificial intelligence in a manner similar to fooling a human being).

availability: Ensuring timely and reliable access to and use of information. SOURCE: SP 800-53; SP 800-53A; SP 800-27; SP 800-60; SP 800-37; FIPS 200; FIPS 199; 44 U.S.C., Sec. 3542

Ayala Scale (Cyber-Physical Attack Severity Levels): A theoretical tool I developed to compare the severity of a cyber-physical attack in the real world, not unlike the Richter scale (earthquakes) and the Fujita scale (tornadoes). A description of the various levels follows.

> **Level 1: Network Probed**
> Denial-of-service, web site defacement, theft of intellectual property. Limited malicious activity. No compromise of services from utilities, communications, or transportation system. No permanent damage to infrastructure equipment or release of toxic materials. No injuries, no fatalities. Minor economic impact to individual or a company.

> **Level 2: Minor Exploit, No Injuries**
> Theft of money, data or intellectual property. Low-risk virus. Temporary disruption of utilities or communications for a building or local transportation system. No permanent damage to infrastructure equipment or release of toxic materials. No injuries, no fatalities. Minor economic impact to a group of individuals, companies, or communities.

Level 3: Light Damage, Some Injuries

Brief localized attack. Disruption of utilities or communications for an individual building, utility company, or citywide transportation system that is repairable. Measured release of non-hazardous materials that can be abated quickly. Superficial injuries, no fatalities. Minor economic impact to a group of individuals, companies, or communities.

Level 4: Moderate Damage, Serious Injuries

Extensive disruption of utilities or communications for multiple buildings or citywide transportation system. Permanent damage to infrastructure equipment for a Regional utility company that will require extensive and expensive repairs. Major equipment must be replaced at great cost and over a long period of time. Measurable release of hazardous materials resulting in serious environmental and collateral damage. Population experiences injuries, no fatalities. Moderate economic impact to a small segment of the population, industries, transportation, and communities.

Level 5: Considerable Damage, Less than 10 Fatalities

Extended disruption of utilities, or communications for multiple buildings, or citywide transportation system. Permanent damage to infrastructure equipment for a Regional utility company that renders equipment unusable and unrepairable. Major equipment replacement required at great cost and over a long period of time. Measurable release of hazardous materials resulting in considerable environmental and collateral damage. Population experiences serious injuries and less than 10 immediate fatalities. Significant economic impact to a large segment of the population, industries, transportation, and communities.

Level 6: Severe Damage, Many Fatalities

Large-scale disruption of Regional utilities, or communications for multiple buildings, or citywide transportation system. Extensive permanent damage to infrastructure equipment for a Regional utility company that renders equipment unusable and unrepairable at multiple sites. Major equipment replacement required at great cost and over a long period of time. Significant release of hazardous materials intended to effect large numbers of casualties and resulting in significant environmental and collateral damage. Population suffers thousands of injuries and hundreds of immediate fatalities. Significant economic impact to a very large segment of the population, industries, transportation, and nation.

Level 7: Devastating Damage

Unlimited and continual cyber-physical attacks intended to effect casualties and/or economic impediments of a nation-state. Unlimited disruption of utilities, or communications for multiple cities or national transportation system. Widespread permanent and irreparable damage to infrastructure equipment for entire cities. Major equipment replacement required at great cost and over a long period of time. Significant release of hazardous materials intended to effect large numbers of casualties and resulting in significant environmental and collateral damage. Population suffers tens of thousands of injuries and loss of life in the thousands. Collapse of a regional economy that affects the entire regional population, industry, utilities, financial sector, transportation sector, and the government.

Level 8: Incredible Damage

Sustained and continual cyber-physical attacks intended to effect casualties and/ or economic impediments of a nation-state. Continued disruption of utilities, or communications for multiple cities or regional transportation systems. Widespread permanent and irreparable damage to infrastructure equipment for entire cities and national utility infrastructure. Major equipment replacement required at great cost and over a long period of time. Extensive release of hazardous materials intended to effect large numbers of casualties and resulting in significant environmental and collateral damage. Population suffers hundreds of thousands of injuries and loss of life in the tens of thousands. Collapse of a national economy that affects the entire population, industries, communications, utilities, financial sector, transportation sector, and the government.

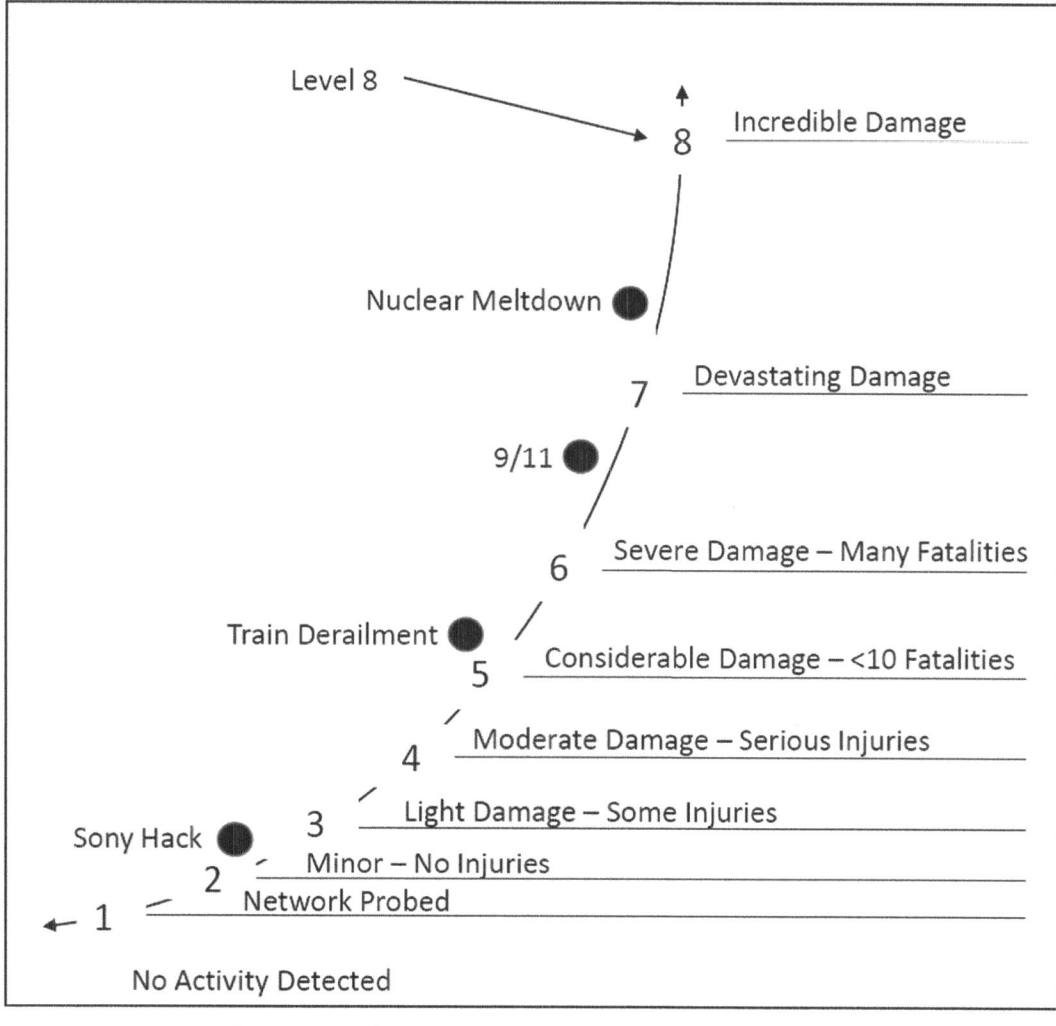

Cyber-Attack Severity Levels (Ayala Scale)

CHAPTER 3

B

baby monitor hacks: Internet-connected baby monitors allow parents to monitor a child's crib. They seldom encrypt the video stream and they can be easily hacked because of inadequate security features. A baby monitor can also be used as an attack vector to penetrate a home's wireless network and monitor other activity. This is a fairly simple hack made possible by finding a device on the same frequency as target device. High-end baby monitors have adjustable frequencies and people tend to use default or simple passwords.

backdoor: Typically, unauthorized hidden software or hardware mechanism used to circumvent security controls. SOURCE: CNSSI-4009

backdoor attack: A *backdoor* is a means of access to a computer program that bypasses security mechanisms. A programmer may sometimes install a backdoor so that the program can be accessed for troubleshooting or other purposes. Also, it can be a hidden method for bypassing building control system authentication. Two types are:

- **beachhead backdoors**: Used to retrieve files, gather building control system information, and trigger execution of other capabilities.

- **standard backdoors**: Communicate using HTTP protocol to blend in with legitimate web traffic or a custom protocol and allow a hacker to upload/download, modify/ delete/execute programs, modify the registry, capture keystrokes, harvest passwords and take screenshots.

back office location: An office or building used to conduct support activities not located within a headquarters.

backblast damper: A passive isolation device installed upstream of a dust collector used to prevent combustion from propagating back upstream in ductwork in the event of a dust collector explosion. The unit contains a flap that is held open by normal airflow. In the event of a pressure wave from an explosion in the collector propagate through the ductwork in the reverse direction of the normal airflow, the backblast damper will force the flap down and prevent the combustion from passing further upstream. Also called an *isolation flap*.

backbone: The underlying network communication conduit or line by which building control system servers and devices are connected; backbone devices are typically servers, routers, hubs, and bridges. Client workstations are not connected directly to the backbone.

backup: Standby equipment for use in the event of the failure of equipment that is normally used, or a copy of data in case the original is damaged. A copy of files and programs made to facilitate recovery, if necessary. SOURCE: SP 800-34; CNSSI-4009

backup agreement: A contract to provide a service as backup that includes the method of performance, the fees, the duration, the services provided, and the extent of security and confidentiality maintained.

backup position listing: A list of alternative personnel who can fill a cyber-physical attack recovery team position when the primary person is unavailable.

backup strategy (recovery strategy): Alternative operating method (i.e., equipment, location, etc.) for building control system operations in the event of a cyber-attack.

backtracking resistance: Backtracking resistance is provided relative to time T if there is assurance that an adversary who has knowledge of the internal state of the Deterministic Random Bit Generator (DRBG) at some time subsequent to time T would be unable to distinguish between observations of ideal random bit strings and (previously unseen) bit strings that were output by the DRBG prior to time T. The complementary assurance is called *prediction resistance*. SOURCE: SP 800-90A

BACnet: The BACnet protocol standard defined by American Society of Heating, Refrigerating and Air Conditioning Engineers (ASHRAE) 135 protocol.

BACnet Interoperability Building Blocks (BIBBs): A BIBB is a collection of one or more BACnet services intended to define a higher level of interoperability. BIBBs are combined to build the BACnet functional requirements for a device in a specification. Some BIBBs define additional requirements (beyond requiring support for specific services) in order to achieve a level of interoperability. For example, the BIBB DS-V-A (Data Sharing-View-A), which would typically be used by a monitoring and controls client, not only requires the client to support the ReadProperty Service, but also provides a list of data types (Object / Properties) that the client must be able to interpret and display.

BACnet network browser: The BACnet network browser is capable of performing full discovery of an ASHRAE 135 building control system including, but not limited to, discovery of all ASHRAE 135 devices, the ASHRAE 135 objects and properties of each device, and the standard ASHRAE 135 services supported by each device. The BACnet network browser is capable of reading any ASHRAE 135 property of any object in any device. Proprietary properties may be presented as read without further interpretation. The BACnet network browser is capable of writing any Standard ASHRAE 135 property of any object in any device. The BACnet network browser also supports segmentation.

bandwidth: The amount of data that can be transmitted via a given communications channel (e.g., between a hard drive and the host PC) in a precise unit of time.

banner: The information displayed to a remote user trying to connect to a service. This may include version information, building control system information, or a warning about unauthorized use. Display on an information system that sets parameters for system or data use. SOURCE: CNSSI-4009

banner grabbing attack: Capturing banner information that is transmitted when a connection is initiated.

beacon channel: A stealthy method to transfer a large amount of information to or from a target network without being detected, because small packets are overlooked by most IDS software. The small packets are then reassembled to a file that can be many megabytes in size. Only deep packet inspection can ferret these out.

baseline: A minimum starting point used for comparisons. Hardware, software, databases, and relevant documentation for an information system at a given point in time. SOURCE: CNSSI-4009

baseline configuration: A set of specifications for a system, or *configuration item* (CI) within a system, that has been formally reviewed and agreed on at a given point in time, and which can be changed only through change control procedures. The baseline configuration is used as a basis for future builds, releases, and/or changes. SOURCE: SP 800-128

baselining: Monitoring building control system to determine typical utilization patterns so that deviations can be detected.

baseline topology: A diagram of the network and network devices as it should be. This is compared to the as-is state of the network to identify any changes that have been made.

basin heater hack: An electric immersion heater installed in a cooling tower to prevent the cold-water basin from completely freezing over during shutdown or standby. It is *not* designed to prevent icing during operation. A hacker can cause the cooling tower to freeze up by shutting down the unit and cutting power to the basin heater. A hacker can also turn the basin heater on in the summer to reduce the efficiency of the unit and run up energy costs.

bastion host: A special-purpose computer on a network specifically designed and configured to withstand attacks. SOURCE: CNSSI-4009

beer pipeline: Bars in the Veltins-Arena, a major multi-function arena in Gelsenkirchen, Germany, are interconnected by a 3.1-mile (5-kilometer) long beer pipeline. A beer pipeline has been proposed for construction in Bruges, Belgium, to reduce truck traffic on the city streets.

behavior monitoring hack: Observing activities of users, building control systems, and processes and measuring the activities against organizational policies and rules, baselines of normal activity, thresholds and trends.

big red button: A *kill switch*, also known as an *emergency stop* or *e-stop*, is a safety mechanism used to shut off a device in an emergency situation in which it cannot be shut down in the usual manner. Unlike a normal shutdown switch/procedure, which shuts down all building control systems in an orderly fashion and turns the machine off without damaging it, a kill switch is designed and configured to (1) completely and as quickly as possible abort the operation, even if this damages equipment; (2) be operable in a manner that is quick, simple (so that even a panicking operator can activate it); and, usually, (3) be obvious even to an untrained operator or a bystander.

binary: A two-state system where the ON condition is represented by a high signal level and the OFF condition is represented by a low signal level. Can also mean a value of 1 or 0 for digital coding of computer systems.

Binary Large OBject (BLOB): A collection of binary data stored as a single entity in a database management system. Blobs are typically images, audio or other multimedia objects, though sometimes binary executable code is stored as a blob. Database support for blobs is not universal. Also might be referred to as the acronym *BLOB* or *data blob*.

BIND: The acronym stands for Berkeley Internet Name Domain, which is an implementation of DNS (Domain Name System). DNS is used for domain name to IP address resolution.

binding: Establishing communications between CEA-709.1-C devices by associating the output of a device to the input of another device so information is automatically sent. Process of associating two related elements of information. SOURCE: SP 800-32

BingDiggity: Bing equivalent of the Google hacking tool.

biohacker: Computer hobbyists who experiment with electronic modifications to the human body. Also called *grinders*. One example is insertion of a glass-encased RFID/NFC chip under the skin of a person's hand that stores information in it that can be read in lieu of a PIN or pattern to unlock a phone. If an office floor requires key fobs, a biohacker can register the identifying information in the chip with building staff and just wave their hand in the elevator, saving the trouble of pulling out a fob. Another example is *embedded fingertip magnets*.

biometric: A measurable physical characteristic or personal behavioral trait used to recognize the identity, or verify the claimed identity, of an applicant. Facial images, fingerprints, and iris scan samples are all examples of biometrics. SOURCE: FIPS 201

BIOS: Basic Input/Output System or Basic Integrated Operating System. The BIOS is installed on a chip in the computer motherboard and has the basic instruction for the operating system to the devices such as video adapter, keyboard mouse, and or hard disk.

bird fancier's lung (BFL): A disease caused by the exposure to avian proteins present in the dry dust of the droppings and sometimes in the feathers of a variety of birds. Bird fancier's lung is a type of hypersensitivity pneumonitis caused by bird droppings. The lungs become inflamed with granuloma formation. Also called *bird-breeder's lung* and *pigeon-breeder's lung*. Someone that has been exposed to avian proteins will see symptoms of BFL within 4 to 6 hours. This condition is occasionally fatal. A building hacker could harvest this material.

birthday attack: A type of cryptographic attack that exploits the mathematics behind the birthday problem in probability theory. This attack can be used to abuse communication between two or more parties because digital signatures can be susceptible to a birthday attack. The attack depends on the higher likelihood of collisions found between random attack attempts and a fixed degree of permutations. Given a function f, the goal of the attack is to find two different inputs—x1, x2—such that f(x1) = f(x2). Such a pair x1, x2 is called a collision. As an example, consider the scenario in which a teacher with a class of 30 students asks for everybody's birthday, to determine whether any two students have the same birthday (corresponding to a hash collision). Intuitively, this chance may seem small. If the teacher picked a specific day (say, September 16), then the chance that at least one student was born on that specific day is about 7.9%. However, the probability that at least one student has the same birthday as any other student is around 70%.

bit-flipping attack: An attack on a cryptographic cipher in which the attacker can change the ciphertext in such a way as to result in a predictable change of the plaintext, although the attacker is not able to learn the plaintext itself. Note that this type of attack is not directly against the cipher itself (as cryptanalysis of it would be), but against a particular message or series of messages. The attack is especially dangerous when the attacker knows the format of the message. In such a situation, the attacker can turn it into a similar message but one in which some important information is altered. For example, a change in the destination address might alter the message route in a way that will force re-encryption with a weaker cipher, thus possibly making it easier for an attacker to decipher the message.

black: Designation applied to encrypted information and the information systems, the associated areas, circuits, components, and equipment processing that information. See also RED. SOURCE: CNSSI-4009

black-box attack: Attacker disconnects an ATM and attaches a computer to command the ATM to dispense cash.

black core: A communication network architecture in which user data traversing a global Internet Protocol (IP) network is end-to-end encrypted at the IP layer. Related to striped core. SOURCE: CNSSI-4009

black hole attack: A type of denial-of-service attack in which a router that is supposed to relay packets instead discards them. This usually occurs from a router becoming compromised from a number of different causes. Also called a *packet drop attack*. One cause mentioned in research is through a denial-of-service attack on the router using a known DDoS tool. Because packets are routinely dropped from a lossy network, the packet drop attack is very hard to detect and prevent. The malicious router can also accomplish this attack selectively; for example, by dropping packets for a particular network destination, at a certain time of the day, a packet every n packets or every t seconds, or a randomly selected portion of the packets. This is rather called a *gray hole attack*. Other information can be found in Request for Comments: 5635 or as Known RFC5635.

black hole e-mail address: An e-mail address which is valid (messages sent to it will not generate errors), but to which all messages sent are automatically deleted, and never stored or seen by humans. These addresses are often used as return addresses for automated e-mails. A hacker may be able to convert a legitimate address to this.

black hole filtering: Black hole filtering refers specifically to dropping packets at the routing level, usually using a routing protocol to implement the filtering on several routers at once, often dynamically to respond quickly to distributed denial-of-service attacks.

black holes (networking): Places in the network where incoming or outgoing traffic is silently discarded (or "dropped"), without informing the source that the data did not reach its intended recipient. When examining the topology of the network, the black holes themselves are invisible, and can only be detected by monitoring the lost traffic; hence the name. The most common form of black hole is simply an IP address that specifies a host machine that is not running or an address to which no host has been assigned. Even though TCP/IP provides means of communicating the delivery failure back to the sender via ICMP, traffic destined for such addresses is often just dropped. Note that a dead address will be undetectable only to protocols that are both connectionless and unreliable (e.g., UDP). Connection-oriented or reliable protocols (TCP, RUDP) will fail to connect to a dead address or will fail to receive expected acknowledgements.

blackers: Jargon for encryption devices that convert red signals to black.

blackhole exploit kit: *Crimeware* sold to hackers that exploits web-browser vulnerabilities.

blacklist: A list of e-mail senders who have previously sent spam to a user. A list of discrete entities, such as hosts or applications that have been previously determined to be associated with malicious activity. SOURCE: SP 800-94 & SP 800-114

blacklisting: The process of the system invalidating a user ID based on the user's inappropriate actions. A blacklisted user ID cannot be used to log on to the system, even with the correct authenticator. Blacklisting and lifting of a blacklisting are both security-relevant events. Blacklisting also applies to blocks placed against IP addresses to prevent inappropriate or unauthorized use of Internet resources. SOURCE: CNSSI-4009

black start hack: The black start is the process of restoring electric power to a building without relying on the commercial power grid. Some power stations have small diesel generators, normally called the *black start diesel generator* (BSDG), which can be used to start larger generators (of several megawatts capacity), which in turn can be used to start the main power station generators. Hacking a black start generator will prevent the large generators from restarting.

black swan event: A metaphor that describes an event that comes as a surprise, has a major effect, and is often inappropriately rationalized after the fact with the benefit of hindsight. The disproportionate role of high-profile, hard-to-predict, and rare events that are beyond the realm of normal expectations in history, science, finance, and technology. The psychological biases that blind people, individually and collectively, to uncertainty and to a rare event's massive role in historical affairs.

blackout test: Shut off commercial power to the building to ensure the generator performs properly.

BladeRF: A software defined radio. Similar to *HackRF* and *USDR*. See *Universal Software Radio Peripheral (USRP)*.

blamestorming: Figuring out who to blame when something goes wrong.

blended threat attack: A hostile action to spread malicious code via multiple methods. For example, sending a malicious URL by e-mail, with text that encourages the recipient to click the link, is a blended threat attack. SOURCE: CNSSI-4009

blinding: Generating network traffic that is likely to trigger many alerts in a short period of time, to conceal alerts triggered by a "real" attack performed simultaneously. SOURCE: SP 800-94

Blippar: Visual search engine using image recognition technology and augmented reality. Available as a free app for Apple and Android devices. Blippar allows a user to use a photograph taken with a smartphone to obtain meaningful information related to the contents of the captured image. For example, a digital photo of a banana would return information found on the web such as what vitamins and how much phosphorous is found in a typical banana. When a user wants to know more about objects, he "blipps" it. Blippar uses Deep Learning (artificial intelligence technology that attempts to simulate the way a human brain works).

block: Sequence of binary bits that comprise the input, output, state, and round key. The length of a sequence is the number of bits it contains. Blocks are also interpreted as arrays of bytes. SOURCE: FIPS 197

block cipher: A symmetric key cryptographic algorithm that transforms one block of information at a time using a cryptographic key. For a block cipher algorithm, the length of the input block is the same as the length of the output block. SOURCE: SP 800-90

block signaling hack: Railway signaling is a system used to safely direct railway traffic in order to prevent trains from colliding. Trains cannot collide with each other if they are not permitted to occupy the same section of track at the same time, so railway lines are divided into sections known as *blocks*. Normally, only one train is permitted in each block at one time. Most blocks are fixed; that is, they include the section of track between two fixed points. Under a moving block system, computers calculate a "safe zone" around each moving train that no other train is allowed to enter. The system depends on knowledge of the precise location and speed and direction of each train, which is determined by a combination of sensors: active and passive markers along the track and trainborne tachometers and speedometers. Hacking the sensors and markers may possibly cause trains to collide or run off the tracks.

blowdown stack hack: A chimney or vertical stack that is used in an emergency to vent the pressure of components of a chemical, refinery, or other process. The purpose is to prevent loss of containment of volatile liquids and gases. Hacking the controls of a blowdown stack could cause pressure to build up. The failure of the blowdown stack to contain vented hydrocarbons led to a catastrophic explosion at a BP refinery in Texas City in 2005.

blue box: A blue box is a tool that emerged in the 1960s and '70s; it allowed users to route their own calls to place free telephone calls. A related device, the black box enabled one to receive calls that were free to the caller. The blue box no longer works as modern switching systems are now digital and do not use in-band signaling.

Blue Screen of Death (BSoD): An error screen displayed on a Windows computer system (Created by Microsoft) after a fatal building control system crash when the operating system reaches a condition where it can no longer operate safely. This is also known as a *stop error*, *blue screen*, or *Blue Screen of Doom*. By default, Windows will create a memory dump file when a stop error occurs.

Blue Team: The group responsible for defending an enterprise's use of information systems by maintaining its security posture against a group of mock attackers (i.e., the Red Team). Typically, the Blue Team and its supporters must defend against real or simulated attacks (1) over a significant period of time, (2) in a representative operational context (e.g., as part of an operational exercise), and (3) according to rules established and monitored with the help of a neutral group refereeing the simulation or exercise (i.e., the White Team). SOURCE: CNSSI-4009

Bluetooth: A wireless technology standard for exchanging data over short distances (typical range 3 to 300 feet) from fixed and mobile devices, and building personal area networks (PANs). Bluetooth is a replacement for cabling in a variety of personally carried applications in any setting, and also works for fixed location applications such as smart energy functionality in the home (thermostats, etc.). It can connect several devices, overcoming problems of synchronization. Bluetooth is a packet-based protocol with a master-slave structure. One master may communicate with up to seven slaves in a piconet. All devices share the master's clock.

Bluesnarfing attack: Hackers gain unauthorized access to the data on a Bluetooth-enabled phone using the wireless technology without alerting the user.

Body of Evidence (BoE): The set of data that documents the information system's adherence to the security controls applied. The BoE will include a Requirements Verification Traceability Matrix (RVTM) delineating where the selected security controls are met and evidence to that fact can be found. The BoE content required by an Authorizing Official will be adjusted according to the impact levels selected. SOURCE: CNSSI-4009

boot sector virus: A virus that plants itself in a system's boot sector and infects the master boot record. SOURCE: SP 800-61

boiler hack: A *boiler* is a closed vessel in which water or other fluid (hydronic) is used to heat a building, whereas a *furnace* uses warm air. The heat is different, and the way that heat is circulated through the building is different. A boiler that has a loss of feed water and is permitted to boil dry can be extremely dangerous. If feed water is then sent into the empty boiler, the small cascade of incoming water instantly boils on contact with the superheated metal boiler shell and leads to a violent explosion that cannot be controlled even by steam safety valves.

boiler lockout: There are two types of lockouts a boiler may experience when a Manual Reset limit device trips, (such as if there is a problem such as low water, high or low pressure, flame failure, blocked flue, or low air). **Manual reset** lockouts require the operator press the reset button. **Automatic reset** lockouts self-reset when the error condition is cleared.

boiler sequence controller hack: *Boiler controller, burner programmer, burner controller, sequence controller*, and *programmable sequence controllers*, measure the temperature of the combined water flow of a multi-boiler installation. They are preprogrammed for the automatic operation of gas/oil burners and regulate how many boilers operate to match the required demand. They continuously monitor the flame and can control how many boilers fire up at one time for safe start-up. They provide output for blower, ignition, and solenoid valves with prefixed timing for continuous flame supervision. Hack this and all the boilers can be turned on full blast.

boot record infector attack: Malware that inserts malicious code into the boot sector of a disk.

booting: The initialization of a computerized building control system. The booting process can be "hard" after electrical power is switched from off to on. "Soft" booting avoiding power-on self-tests can be initiated by hardware such as a button press, or by software command. Booting is complete when normal operation is attained.

Border Gateway Protocol (BGP): An inter-autonomous building control system routing protocol. BGP is used to exchange routing information for the Internet between Internet Service Providers (ISPs).

bot attack: An application that runs automated cyber-attacks over the Internet. Bots perform simple and repetitive tasks at a faster rate than humans can.

bot master: The controller of a botnet that directs compromised computers in the botnet from a remote location.

botnet attack: A group of computers taken over by malicious software and controlled across a network. The compromised computers are commonly known as *zombies*. These computers, which have been infected with malware, allow the attacker to control them.

boundary protection: Monitoring and control of communications at the external boundary of an information system to prevent and detect malicious and other unauthorized communication, through the use of boundary protection devices (e.g., proxies, gateways, routers, firewalls, guards, encrypted tunnels). SOURCE: SP 800-53; CNSSI-4009

boundary protection device: A device with appropriate mechanisms that (1) facilitates the adjudication of different interconnected system security policies (e.g., controlling the flow of information into or out of an interconnected system); and/or (2) provides information system boundary protection. SOURCE: SP 800-53 A device with appropriate mechanisms that facilitates the adjudication of different security policies for interconnected systems. SOURCE: CNSSI-4009

brittle system: A building control system characterized by a sudden and steep decline in performance due to input parameters that exceed a specified input, or environmental conditions that exceed specified operating boundaries. The opposite of a gracefully degrading building control system.

bridge: A product that connects one local area network to another that uses the same protocol (for example, token ring or Ethernet).

bricked: When an intelligent electronic device such as a router or computer can no longer function, due to a serious misconfiguration, corrupted firmware, or a hardware problem. The device is as technologically useful as a brick. Bricking suggests that the damage is so serious as to have rendered the intelligent electronic device permanently unusable. Devices can be bricked by malware and sometimes by running software not intentionally harmful, but with errors that cause damage. Some devices include two copies of firmware, one active and the other stored in fixed ROM or writable non-volatile memory and not normally accessible to processes that could corrupt it, as well as a way to copy the stored firmware over the active version, even if corrupt, so that if the active firmware is damaged, it can be replaced by the copy and the device will not be bricked. Bricking is classified into two types— namely soft brick and hard brick, depending on the device's ability to function.

- **Soft-bricked** devices generally show some signs of life. A soft bricked device usually boots unsuccessfully and generally gets stuck on boot logo, or reboots endlessly, or suddenly shows a "screen of death." Some devices can recover from a soft-bricked state by simply clearing all the internal memory and flashing the firmware.

- **Hard-bricked** devices show little to no signs of life. A hard-bricked device doesn't power up or show a vendor logo, basically the screen remains turned off. Recovering from a hard brick is difficult and requires the use of a direct programming interface to the controller. Sometimes referred to as a *doorstop*.

broadcast: Unlike most messages, which are intended for a specific recipient device, a broadcast message is intended for all devices on the network without acknowledgement by the receivers. SOURCE: UFGS-25 10 10

browsing: Act of searching through information system storage or active content to locate or acquire information, without necessarily knowing the existence or format of information being sought. SOURCE: CNSSI-4009

brute-force attack: A method of accessing an obstructed device through attempting multiple combinations of numeric and/or alphanumeric passwords. An attacker tries to use all possible combinations of letters, numbers, and symbols to enter a correct password. Programs exist to do this, such as Zip Password Cracker Pro. Any password can be cracked using the brute-force method, but it can take a very long time so this is most popular in movies, and less so in real life. The longer and more intricate a password is, the longer it will take a computer to try all of the possible combinations. SOURCE: SP 800-72

BTU meter hack: A BTU meter provides highly accurate thermal energy measurement in chilled water, hot water, and condenser water systems based on signal inputs from temperature sensors and any of insertion or inline flow meters. The LonWorks model provides energy, flow, and temperature data to the LonTalk network using standard network variable types (SNVT). A local indication of energy, flow, and temperature data is also available through an alphanumeric display located on the BTU meter. An optional auxiliary input is also available to totalize pulses from another device and communicate the total directly to the network. If a hacker can access the BTU meter he can send erroneous signals to the BCS and spoof the controls so maintenance staff is unaware of the problem.

buffer overflow: A condition at an interface under which more input can be placed into a buffer or data holding area than the capacity allocated, overwriting other information. Attackers exploit such a condition to crash a system or to insert specially crafted code that allows them to gain control of the system. SOURCE: SP 800-28; CNSSI-4009

buffer overflow attack: A method of overloading a predefined amount of space in a buffer, which can potentially overwrite and corrupt data in memory. Hackers exploit such a condition to crash a building control system or to insert specially crafted code that allows them to gain control of the building control system. SOURCE: SP 800-28; CNSSI-4009

bugcrowd: Crowdsourced bug-hunting middlemen for friendly hackers who find vulnerabilities and disclose them to the company for a reward so that they can get fixed. Other bugcrowd companies include Crowdcurity, Synack, and HackerOne. HackerOne handles the legal and logistical details, taking care of billing and payment in exchange for a 20% commission on top of each bounty.

Bugtraq: A mailing list for the discussion of security problems and vulnerabilities. Occasionally, full disclosure reports of new vulnerabilities and exploit tools are distributed through this list.

buildering hack: The act of climbing on the outside of buildings and other structures, mostly undertaken at nighttime by college students. The word is a portmanteau, combining the word building with the climbing term bouldering (also known as *edificeering*, *urban climbing*, *structuring*, or *stegophily*).

building automation system (BAS): A computer-based control system installed in buildings that controls and monitors the building's mechanical and electrical equipment such as ventilation, lighting, power systems, fire systems and security systems. A BAS consists of software and hardware; the software program, usually configured in a hierarchical manner, can be proprietary, using such protocols as C-bus, Profibus, and so on. Vendors are also producing BASs that integrate using Internet protocols and open standards such as DeviceNet, SOAP, XML, BACnet, LonWorks, and Modbus.

building control network (BCN): The network that connects devices used by the building control system. Typically, the BCN is a BACnet ASHRAE 135 or LonWorks CEA-709.1-D network installed by the building control system subcontractor. SOURCE: UFGS-25 10 10

building control system (BCS): A control system for building electrical and mechanical systems, HVAC (including central mechanical plants) and lighting. A BCS generally uses DDC hardware and generally does *not* include its own local front end. SOURCE: UFGS-25 10 10

building hack: A building hack can be something as simple as an employee unlocking a thermostat cover to change the office temperature to a hacker breaking into a building control system and using a computer program to rapidly open and close a diesel generator's circuit breakers out of phase from the rest of the grid causing it to explode. This vulnerability is referred to as the *Aurora vulnerability*.

building HVAC zones: The floor areas of a building served by an HVAC system is divided into zones. Each zone can be set to unique temperature control such as the building perimeter zones and open office areas in the center of a building. A hacker with access to the BCS can cool the building perimeter zones in the winter and heat the building perimeter in the summer (opposite of proper operations).

building impact analysis (BIA): The process of analyzing all building functions and the effect that a specific disaster or cyber-attack may have upon them. SOURCE: NICCS

building interruption hacks: An event that disrupts the normal building operations at a specific facility.

building interruption costs: The cost associated with an interruption in normal building operations.

building information modeling (BIM): The process involving the generation and management of digital representations of physical and functional characteristics of places. BIMs are files (often but not always in proprietary formats and containing proprietary data) which can be exchanged or networked to support decision-making about a facility. Current BIM software is used by individuals, businesses and government agencies who plan, design, construct, operate and maintain diverse physical infrastructures, such as water, wastewater, electricity, gas, refuse and communication utilities, roads, bridges and ports, houses, apartments, schools and shops, offices, factories, warehouses and prisons. BIM Washing is a term describing the inflated, and/or deceptive, claim of using or delivering Building Information Modeling services or products. SOURCE: Wikipedia

Building Internet of Things (BIoT) hacks: As buildings become increasingly connected, the attack surface expands exponentially. As more and more devices are connected to the Internet, society becomes more and more vulnerable to hackers. Everything from baby monitors to garage doors can be hacked with unintended consequences. Hackers have even demonstrated the ability to unlock prison cell doors in a high security prison. The BIoT can become a cheap tool for remote surveillance and reconnaissance and a great deal of information can be learned from device usage behavior. By aggregating this surveillance information about your building over time, an attacker could get a very accurate understanding of your building operations.

Internet tools such as Wink allow a person to control, from a single screen, his Internet-connected home devices, such as door locks, window shades, and LED lights. Anyone capable of hacking your Wink account may be able to identify your social media accounts, the names of your devices (like Lou's iPad) and your network information. An app that monitors your grill's propane tank may record the tank's latitude and longitude, thus revealing the exact location of your house. Hacking into a *Nest* thermostat would allow someone to figure out when your house was occupied and when it was not. Manufacturers of IP-enabled devices say you can opt out of sharing information with vendors, software developers and third-party applications, but you may not be aware how much information their device is collecting.

building management system (BMS): Same as *building automation system*.

building operations recovery: The component of recovery after a cyber-physical attack which deals specifically with the relocation of key personnel, provision of equipment, supplies, work space, communication facilities, computer processing capability, records and so forth.

building point of connection (BPOC): (1) The place where the utility company connects to the building, which determines who is responsible for installation and maintenance of the equipment: the building owner or the utility company. The demarcation point varies with the type of utility and also varies by country. (2) The point of hardware connection between the UMCS network backbone and the building control network.

building setback: The distance from the outside perimeter of a government installation to the building.

building recovery coordinator: See *disaster recovery coordinator.*

building recovery process: The critical path of events that need to be followed after a cyber-physical attack.

building recovery team: Individuals responsible for the building recovery process.

building resumption planning (BRP): The operations piece of building recovery planning.

bump test: Check of basic functionality of a gas detector or monitor by exposure to test gas, resulting in a gas indication or alarm condition.

burner management system: Controls and operates various applications in the oilfield. Critical information including temperature, pressure, and flame detection is managed to facilitate a safe combustion environment. SOURCE: Profire Energy Inc.

burp suite: A scanner with a limited "intruder" tool for cyber-attacks. Many security-testing specialists swear that pen testing without this tool is unimaginable. Very effective.

burst mode: A quick high-speed data transfer at significantly higher rates than would normally be achieved and the maximum throughput a device is capable of.

bus: The main electrical communication path in which signals are sent from one part of the computer to another.

business continuity planning (BCP): An umbrella term covering both disaster recovery planning and building resumption planning.

butterfly effect: In chaos theory, it is the sensitive dependence on initial conditions in which a small change in one state of a deterministic nonlinear system can result in large differences in a later state. If the theory were correct, one flap of a seagull's wings would be enough to alter the course of the weather.

byte: The fundamental data unit for personal computers, 8 contiguous bits.

Byzantine fault: A fault presenting different symptoms to different observers.

Byzantine fault tolerance (BFT): The objective of BFT is to be able to provide the building control system's service assuming there are not too many faulty components. Some aircraft systems, such as the Boeing 777 Aircraft Information Management System and the Boeing 787 flight control systems, use Byzantine fault tolerance.

Byzantine failure hack: The loss of a building control system service due to a Byzantine fault in systems that require consensus. Failure occurs when components of a building control system fail with symptoms that prevent some components of the building control system from reaching agreement among themselves, where such agreement is needed for the correct operation of the building control system.

CHAPTER 4

C

cache: A random access memory component used for temporary storage of information so future requests for that information can be retrieved faster.

cache cramming attack: The technique of tricking a computer browser to run cached Java code from the local disk, instead of the Internet zone, so it runs with less restrictive permissions.

cache poisoning attack: Bad data from a remote name server is cached by another name server. Typically used with DNS cyber-attacks.

cache stampede attack: A type of cascading failure that can occur when building control systems with caching mechanisms come under very high load. Sometimes also called *dog-piling*.

Cain & Abel: A tool for cracking encrypted passwords or network keys. It uses network sniffing, Dictionary, Brute-Force, and Cryptanalysis attacks, cache uncovering and routing protocol analysis methods.

Call Admission Control (CAC): Inspection of inbound and outbound voice network activity by a voice firewall based on user-defined policies.

call back: A procedure for identifying a remote terminal. In a call back, the host building control system disconnects the caller and then dials the authorized telephone number of the remote terminal to reestablish the connection. Synonymous with *dial-back*. SOURCE: CNSSI-4009

call back attack: One of the first things a hacker does when he gets into a BCS is install a routine with automatic call back in case the system is disconnected so he can reestablish connection when system is rebooted.

canary: Anything that can send up an observable alert if something happens. For example, you can set up a computer on a subnet such that no other computer should ever access that. If something touches it, you know it's from outside normal behavior. Also called a *tripwire*.

cantenna: A portmanteau blending the words can and antenna is a homemade directional waveguide antenna, made out of an open-ended metal can. Cantennas are typically used to increase the range of (or discover) Wi-Fi networks. Although some designs are based on a Pringles potato chips can, this tube is too narrow to increase the 2.4 GHz signal by a useful amount, although at 5 GHz it would be about the right size.

cascading: Downward flow of information through a range of security levels greater than the accreditation range of a system, network, or component. SOURCE: CNSSI-4009

cascading failure hack: A failure in a building control system of interconnected parts in which the failure of one part can trigger the failure of successive parts. When one part of the building control system fails, nearby nodes must take up the slack for the failed component. This overloads these nodes, causing them to fail as well, prompting additional nodes to fail one after another. Electrical blackouts frequently result from a cascade of failures between interdependent networks.

© Luis Ayala 2016
L. Ayala, *Cybersecurity Lexicon*, DOI 10.1007/978-1-4842-2068-9_4

cascading rollback: Occurs in database systems when a transaction (T1) causes a failure and a rollback must be performed. Other transactions dependent on T1's actions must also be rolled backed due to T1's failure, thus causing a cascading effect.

catfish: Someone who pretends to be someone else using social media to create a false identity, typically to pursue online romances.

cell: A cell is a unit of data transmitted over an ATM network. ATM cell consists of a 5-byte header and a 48-byte payload.

central office: A secure, self-contained telecommunications equipment building that houses servers, storage systems, switching equipment, emergency power systems, and related devices that are used to run telephone systems.

central mechanical plant (CMP): A building (or large room) dedicated to the mechanical equipment and electrical equipment necessary for a large building or campus. A CMP typically houses air handlers, boilers, chillers, heat exchangers, water heaters, water pumps (for domestic, heating/cooling, and firefighting water), main distribution piping and valves, sprinkler distribution piping and pumps, back-up electrical generators, elevator machinery, and other HVAC equipment.

Central Services Node (CSN): The Key Management Infrastructure core node that provides central security management and data management services. SOURCE: CNSSI-4009

central utility plant (CUP): Same as *central mechanical plant.*

certificate: A digital representation of information which at least

- identifies the certification authority issuing it,

- names or identifies its subscriber,

- contains the subscriber's public key,

- identifies its operational period, and

- is digitally signed by the certification authority issuing it.

SOURCE: SP 800-32

certificate policy (CP): A specialized form of administrative policy tuned to electronic transactions performed during certificate management. A certificate policy addresses all aspects associated with the generation, production, distribution, accounting, compromise recovery, and administration of digital certificates. Indirectly, a certificate policy can also govern the transactions conducted using a communications system protected by a certificate-based security system. By controlling critical certificate extensions, such policies and associated enforcement technology can support provision of the security services required by particular applications. SOURCE: CNSSI-4009; SP 800-32

Certificate Revocation List (CRL): A list of revoked public key certificates created and digitally signed by a certification authority. SOURCE: SP 800-63; FIPS 201

certification: A comprehensive assessment of the management, operational, and technical security controls in a building control system, made in support of security accreditation, to determine the extent to which the controls are implemented correctly, operating as intended, and producing the desired outcome with respect to meeting the security requirements for the building control system. SOURCE: FIPS 200

certification analyst: The independent technical liaison for all stakeholders involved in the Certification & Accreditation process responsible for objectively and independently evaluating a system as part of the risk management process. Based on the security requirements documented in the security plan, performs a technical and non-technical review of potential vulnerabilities in the system, and determines if the security controls (management, operational, and technical) are correctly implemented and effective. SOURCE: CNSSI-4009

certified disaster recovery planner (CDRP): Individual certified by the Disaster Recovery Institute.

certifier: Individual responsible for making a technical judgment of the system's compliance with stated requirements, identifying and assessing the risks associated with operating the system, coordinating the certification activities, and consolidating the final certification and accreditation packages. SOURCE: CNSSI-4009

chemical feed pump: A relay- or proportionally-controlled pump that dispenses a chemical.

chain/loop attack: A chain of connections through many nodes as the attacker moves across multiple nodes to hide his origin and identity. In case of a loop attack, the chain of connections is in a loop making it more difficult to track down his origin.

chain of custody: A process that tracks the movement of evidence through its collection, safeguarding, and analysis lifecycle by documenting each person who handled the evidence, the date/time it was collected or transferred, and the purpose for the transfer. SOURCE: SP 800-72; CNSSI-4009

chain of evidence: A process and record that shows who obtained the evidence; where and when the evidence was obtained; who secured the evidence; and who had control or possession of the evidence. The "sequencing" of the chain of evidence follows this order: collection and identification; analysis; storage; preservation; presentation in court; return to owner. SOURCE: CNSSI-4009

challenge and reply authentication: Prearranged procedure in which a subject requests authentication of another and the latter establishes validity with a correct reply. SOURCE: CNSSI-4009

Challenge-Handshake Authentication Protocol: An authentication protocol where the verifier sends the claimant a challenge (usually a random value or a nonce) that the claimant combines with a shared secret (often by hashing the challenge and secret together) to generate a response that is sent to the verifier. The verifier knows the shared secret and can independently compute the response and compare it with the response generated by the claimant. If the two are the same, the claimant is considered to have successfully authenticated himself. When the shared secret is a cryptographic key, such protocols are generally secure against eavesdroppers. When the shared secret is a password, an eavesdropper does not directly intercept the password itself, but the eavesdropper may be able to find the password with an off-line password guessing attack. SOURCE: SP 800-63

channel: A portion of the control network consisting of one or more segments connected by repeaters. SOURCE: Army Corps Spec 15951

Character Generator Protocol (CHARGEN): A service of the Internet Protocol Suite intended for testing, debugging, and measurement purposes that is rarely used, as its design flaws allow misuse. UDP CHARGEN is commonly used in denial-of-service attacks. By using a fake source address, the attacker can send bounce traffic off a UDP CHARGEN application to the victim. UDP CHARGEN sends 200 to 1,000 times more data than it receives, depending upon the implementation. This "traffic multiplication" is attractive to an attacker. Also attractive is the obscuring of the attacker's IP address from the victim. CHARGEN was widely implemented on network-connected printers. As printer firmware was rarely updated on older models before CHARGEN and other security concerns were known, there may still be many network-connected printers that implement the protocol. Where these are visible to the Internet, they are invariably misused as denial-of-service vectors. **Potential attackers often scan networks looking for UDP port 19 CHARGEN sources.** So notorious is the availability of CHARGEN in printers that some distributed denial-of-service Trojans now use UDP port 19 for their attack traffic. The supposed aim is to throw investigators off the track; to have them looking for old printers rather than subverted computers. Another source: RFC 864 by Jon Postel.

checklist test: A method used to ensure accuracy of information in a disaster recovery plan and that it is current.

checksum: Value computed on data to detect error or manipulation. SOURCE: CNSSI-4009

check word: Cipher text generated by cryptographic logic to detect failures in cryptography. SOURCE: CNSSI-4009

chilled beam: A chilled beam is a type of convection HVAC system designed to heat or cool large buildings. Pipes of water are passed through a *beam* (a *heat exchanger*) either integrated into a suspended ceiling system or suspended a short distance from the ceiling. Not a structural member.

chilled water loop: Chilled water systems are the primary means of cooling commercial and industrial facilities. Chilled water systems transfer heat from water pumped through a piping network to a refrigerant fluid. A closed loop system uses water as the cooling media that does not exit or enter the system once in operation. When a chiller is connected through piping to the load, chilled water can be pumped throughout the network to manage loads in the primary loop as well as a secondary loop.

chiller hack: Four kinds of water chillers are used for buildings and industrial use. The most widely used: centrifugal: are found in large industrial and commercial buildings. Other types (reciprocating, rotary screw, and absorption chillers) are found in small to mid-sized facilities operations. Chillers are controlled by the BCS so if the BCS is hacked an attacker could break them such as by making them surge by altering CHW/CW setpoints. Gas fired absorption chillers are potentially more dangerous, but all chillers have internal safeties to shut them down for a variety of reasons. Taking the chiller plant off line is probably the biggest danger.

chiller usage report: A report of the operation of each chiller as shown on a daily and monthly basis, for each of at least 10 discrete loading levels. The report includes:

- Average power for the month at each level in kW.

- Total monthly energy use in kWh at each level.

- Total monthly energy use in kWh for the chiller (all levels).

- Total daily run hours at each level.

- Total Monthly run hours at each level.

churn test: A weekly test that verifies the operational status of a building's fire pump and identifies deficiencies that may not be evident by visual examination.

Cinderella attack: A cyber-attack that disables security software by manipulating the network internal clock time so a security software license expires prematurely rendering the target network vulnerable to cyber-attack.

ciphony: Process of enciphering audio information, resulting in encrypted speech.

circuit breaker: An automatically operated electrical switch designed to protect an electrical circuit from damage caused by overload or short circuit. Its basic function is to detect a fault condition and interrupt current flow. Unlike a fuse, which operates once and then must be replaced, a circuit breaker can be reset (either manually or automatically) to resume normal operation. The following are various types of circuit breakers:

- air break circuit breaker

- oil circuit breaker (tank type of bulk oil)

- minimum oil circuit breaker

- air blast circuit breaker

- vacuum circuit breaker

- sulphur hexafluoride circuit breaker

circuit switched network: Circuit switching is a methodology of implementing a telecommunications network in which two network nodes establish a dedicated communication channel (circuit) through the network before the nodes may communicate. The circuit guarantees the full bandwidth of the channel and remains connected for the duration of the communication session. The circuit functions as if the nodes were physically connected as with an electrical circuit.

City Tie Module (CTM): The connection to the local fire department or a central monitoring agency to speed response time. Also known as the *fire department relay*.

clear: To use software or hardware products to overwrite storage space on the media with nonsensitive data. This process may include overwriting not only the logical storage location of a file(s) (e.g., file allocation table) but also may include all addressable locations. SOURCE: SP 800-88

clear text: Information that is not encrypted. SOURCE: SP 800-82

cleared employee: All employees granted PCLs and all employees being processed for PCLs.

clearing: Removal of data from an information system, its storage devices, and other peripheral devices with storage capacity, in such a way that the data may not be reconstructed using common system capabilities (i.e., through the keyboard); however, the data may be reconstructed using laboratory methods. SOURCE: CNSSI-4009

clickbait: A pejorative term describing web content that is aimed at generating online advertising revenue, especially at the expense of quality or accuracy, relying on sensationalist headlines to attract click-throughs and to encourage forwarding of the material over online social networks. Clickbait headlines typically aim to exploit the *curiosity gap*, providing just enough information to make the reader curious, but not enough to satisfy their curiosity without clicking through to the linked content.

click farm: A form of click fraud, where a large group of low-paid workers are hired to click on paid advertising links for the click fraudster (click farm master or click farmer). The workers click the links, surf the target web site for a period of time, and possibly sign up for newsletters prior to clicking another link. For many of these workers, clicking on enough ads per day may increase their revenue substantially and may also be an alternative to other types of work. It is extremely difficult for an automated filter to detect this simulated traffic as fake because the visitor behavior appears exactly the same as that of an actual legitimate visitor. Although click farms services violates social media user policy, there is no government regulations that render them illegal.

click-jacking: Concealing hyperlinks beneath legitimate clickable content which, when clicked, causes a user to unknowingly perform actions, such as downloading malware, or sending your ID to a site. Numerous click-jacking scams have employed Like and Share buttons on social networking sites. Disable scripting and iframes in whatever Internet browser you use. SOURCE: FBI Internet Social Networking Risks

client: A building control system computer that requests and uses a service provided by a "server" computer. Sometimes the server may be a client of some other server.

cloud computing: A model for enabling on-demand network access to a shared pool of configurable IT capabilities/ resources (e.g., networks, servers, storage, applications, and services) that can be rapidly provisioned, and released with minimal management effort or service provider interaction. It allows users to access technology-based services from the network cloud without knowledge of, expertise with, or control over the technology infrastructure that supports them. This cloud model is composed of five essential characteristics (on-demand self-service, ubiquitous network access, location independent resource pooling, rapid elasticity, and measured service); three service delivery models (*Cloud Software as a Service* [SaaS], *Cloud Platform as a Service* [PaaS], and *Cloud Infrastructure as a Service* [IaaS]); and four models for enterprise access (*private cloud, community cloud, public cloud,* and *hybrid cloud*). SOURCE: CNSSI-4009

cloud video recording (CVR): Installation of an Internet-connected digital camera that stores recorded video that can be accessed from anywhere.

clustered servers: Connecting many computers together through an Ethernet backbone, to form a ring of host computers that acts as a single entity, distributing the workload across all members of the ring to perform complex operations.

clustered storage: Connecting many data storage computers to form a redundant ring of data storage devices capable of performing multiple read and write requests to the requesting computer.

clutron: Cracker-speak for an indivisible particle of cluefulness. Also referred to as a *cluon.*

cogeneration plant: Cogeneration, or combined heat and power (CHP), is the use of a power station to generate electricity and capture waste heat simultaneously.

Cognizant Security Agency (CSA): Agencies of the executive branch that have been authorized to establish an industrial security program to safeguard classified information under the jurisdiction of those agencies when disclosed or released to US industry. These agencies are the US Department of Defense, the US Department of Energy, the Central Intelligence Agency (CIA), and the Nuclear Regulatory Commission (NRC). SOURCE: Executive Order 12829

Cognizant Security Office (CSO): The organizational entity delegated by the head of a CSA to administer industrial security on behalf of the CSA. SOURCE: 32 CFR Part 2004

cold site: Backup site that can be up and operational in a relatively short time span, such as a day or two. Provision of services, such as telephone lines and power, is taken care of, and the basic office furniture might be in place, but there is unlikely to be any computer equipment, even though the building might well have a network infrastructure and a room ready to act as a server room. In most cases, cold sites provide the physical location and basic services. SOURCE: CNSSI-4009

collateral effect: Unintentional or incidental effects including, but not limited to, injury or damage to persons or objects that would not be lawful military targets under the circumstances ruling at the time. Includes effects on civilian or dual-use computers, networks, information, or infrastructure. Such effects are not unlawful as long as they are not excessive in light of the overall military advantage anticipated from the activity. In cyberspace operations, collateral effects are categorized as:

- **High:** Substantial adverse effects on persons or property that are not lawful targets from which there is a reasonable probability of loss of life, serious injury, or serious adverse effect on the affected nation's security, economic security, public safety, or any combination of such effects.

- **Medium:** Substantial adverse effects on persons or property that are not lawful targets.

- **Low:** Temporary, minimal or intermittent effects on persons or property that are not lawful targets.

- **No:** Only adversary persons and computers, computer-controlled networks, and/or information and information systems are adversely affected.

collision: Two or more distinct inputs produce the same output. Also see *Birthday Attack* for another definition of collision. SOURCE: SP 800-57

collision attack: In cryptography, a collision attack on a cryptographic hash tries to find two inputs producing the same hash value, such as a hash collision. See *Birthday Attack*.

commercial and government entity code (CAGE): DSS uses CAGE Codes to track basic facility information.

command and/or control center (CAC/CNC/CCC): A central facility with adequate telephone capacity to coordinate the recovery process until the site incapacitated by a cyber-attack is functional.

commandable: A point (Object) is commandable if its Present_Value Property is writable and it supports the optional Priority_Array Property. This functionality is useful for Overrides. SOURCE: UFGS-25 10 10

comment ghosting: In online community, where an individual comment is rendered invisible to everyone except the poster in order to eliminate disruption they might otherwise cause. Also referred to as *stealth banning* or *ghost-posting*.

Commerce Service Provider (CSP): A company that provides e-commerce solutions.

commercial off-the-shelf (COTS): Commercially available products that can be purchased and integrated with little customization, facilitating customer installation and reducing costs.

commissioning: The process of assuring that all systems and components of a building or industrial plant are designed, installed, tested, operated, and maintained properly for optimum performance. The commissioning process is typically performed by a third party subcontractor and may be applied to new construction or renovation projects. Written reports are submitted as a record of the proper operation and these can be referred to later on when equipment doesn't seem to be running properly.

commodity service: An information system service (e.g., telecommunications service) provided by a commercial service provider typically to a large and diverse set of consumers. The organization acquiring and/or receiving the commodity service possesses limited visibility into the management structure and operations of the provider, and while the organization may be able to negotiate service-level agreements, the organization is typically not in a position to require that the provider implement specific security controls. SOURCE: SP 800-53

common access card (CAC): Standard identification/smart card issued by the Department of Defense that has an embedded integrated chip storing public key infrastructure (PKI) certificates. SOURCE: CNSSI-4009

Common Gateway Interface (CGI): Used by web servers to pass parameters to executable scripts to generate responses dynamically.

Common Industrial Protocol (CIP): An industrial protocol for industrial automation applications. CIP encompasses a comprehensive suite of messages and services for the collection of manufacturing automation applications: control, safety, synchronization, motion, configuration, and information. CIP allows users to integrate these manufacturing applications with enterprise-level Ethernet networks and the Internet. CIP is media-independent and provides a unified communication architecture throughout the manufacturing enterprise. These include application extensions to CIP: CIP Safety, CIP Motion, and CIP Sync. SOURCE: Wikipedia

Common Misuse Scoring System (CMSS): A set of measures of the severity of software feature misuse vulnerabilities. A software feature is a functional capability provided by software. A software feature misuse vulnerability is a vulnerability in which the feature also provides an avenue to compromise the security of a building control system. SOURCE: NIST Interagency Report (IR) 7864

Common Vulnerabilities and Exposures (CVE): A dictionary of common names for publicly known information system vulnerabilities. SOURCE: SP 800-51; CNSSI-4009

Common Vulnerability Scoring System (CVSS): A SCAP specification for communicating the characteristics of vulnerabilities and measuring their relative severity. SOURCE: SP 800-53A

communications cover: Concealing or altering of characteristic communications patterns to hide information that could be of value to an adversary. SOURCE: CNSSI-4009

communications deception: Deliberate transmission, retransmission, or alteration of communications to mislead an adversary's interpretation of the communications. SOURCE: CNSSI-4009

communications profile: Analytic model of communications associated with an organization or activity. The model is prepared from a systematic examination of communications content and patterns, the functions they reflect, and the communications security measures applied. SOURCE: CNSSI-4009

communications failure: An unplanned interruption in electronic communication as a result of hardware or software failure.

communications recovery: The restoration or rerouting of a communication network, or its components after a cyber-physical attack.

compartmented mode: Mode of operation wherein each user with direct or indirect access to a system, its peripherals, remote terminals, or remote hosts has all of the following: (1) valid security clearance for the most restricted information processed in the system; (2) formal access approval and signed nondisclosure agreements for that information which a user is to have access; and (3) valid need-to-know for information which a user is to have access. SOURCE: CNSSI-4009

competitive intelligence: Legal, or at least not obviously illegal, espionage.

Competitive Local Exchange Carrier (CLEC): A telecommunications provider company (sometimes called a *carrier*) competing with other, already established carriers (generally the incumbent local exchange carrier (ILEC)). Most CLECs also offer Internet services.

comprehensive testing: A test methodology that assumes explicit and substantial knowledge of the internal structure and implementation detail of the assessment object. Also known as *white box testing*. SOURCE: SP 800-53A

compromise: Disclosure of information to unauthorized persons, or a violation of the security policy of a system in which unauthorized intentional or unintentional disclosure, modification, destruction, or loss of an object may have occurred. SOURCE: SP 800-32

compromising emanations: Unintentional signals that, if intercepted and analyzed, would disclose the information transmitted, received, handled, or otherwise processed by information systems equipment. SOURCE: CNSSI-4009

computer-aided design (CAD): The use of a computer in design applications such as architecture, engineering, and manufacturing.

computer-aided facility management (CAFM): Software and systems that enable facility managers to increase the utilization of space and facilities, reduce office moves and reallocations, plan preventative maintenance, efficiently execute reactive maintenance, standardize services, and streamline processes. Ultimately, information from CAFM software allows managers to improve long term planning of real estate, space, facilities, maintenance, and service requirements against budgets to ensure alignment with core business needs. SOURCE: Wikipedia

computer forensics: The practice of gathering, retaining, and analyzing computer-related data for investigative purposes in a manner that maintains the integrity of the data. SOURCE: SP 800-61; CNSSI-4009

Computer Incident Response Team (CIRT): Group of individuals usually consisting of Security Analysts organized to develop, recommend, and coordinate immediate mitigation actions for containment, eradication, and recovery resulting from computer security incidents. Also called a *Computer Security Incident Response Team* (CSIRT) or a *CIRC* (*Computer Incident Response Center, Computer Incident Response Capability, or Cyber Incident Response Team*). SOURCE: CNSSI-4009

computer network attack (CNA): Actions taken through the use of computer networks to disrupt, deny, degrade, or destroy information resident in computers and computer networks, or the computers and networks themselves. A category of *fires* employed for offensive purposes in which actions are taken through the use of computer networks to disrupt, deny, degrade, manipulate, or destroy information resident in the target information system or computer networks, or the systems/networks themselves. The ultimate intended effect is not necessarily on the target system itself, but may support a larger effort, such as information operations or counter-terrorism; for example, altering or spoofing specific communications or gaining or denying access to adversary communications or logistics channels. The term *fires* means the use of weapon systems to create specific lethal or nonlethal effects on a target. SOURCE: CNSSI-4009

Computer Network Defense (CND): Actions taken to defend against unauthorized activity within computer networks. CND includes monitoring, detection, analysis (such as trend and pattern analysis), and response and restoration activities. SOURCE: CNSSI-4009

Computer Network Exploitation (CNE): Enabling operations and intelligence collection capabilities conducted through the use of computer networks to gather data from target or adversary information systems or networks. SOURCE: CNSSI-4009

Computer Network Defense Analysis: In cybersecurity work, a person that uses defensive measures and information collected from a variety of sources to identify, analyze, and report events that occur or might occur within the network in order to protect information, building control systems, and networks from threats. SOURCE: NICCS

Computer Network Defense Infrastructure Support: In cybersecurity work, a person that tests, implements, deploys, maintains, reviews, and administers the infrastructure hardware and software that are required to effectively manage the computer network defense service provider network and resources; monitors network to actively remediate unauthorized activities. SOURCE: NICCS

Computer Network Operations (CNO): Comprised of computer network attack, computer network defense, and related computer network exploitation enabling operations. SOURCE: CNSSI-4009

Computer Numerical Control (CNC): A machine tool whose function and motions are controlled by a software program. CNC can control the as feed, depth of cut, speed, and the functions such as turning spindle on/off, turning coolant on/off.

Computer Recovery Team (CRT): The staff responsible for assessing damage to the building control system, restoring building equipment operation, and setting up a new building control system.

Computer Room Air Conditioner (CRAC): A dedicated air conditioner that monitors and maintains the temperature, air distribution, and humidity in a network room or data center.

Computer Security Incident Response Team (CSIRT): A capability set up for the purpose of assisting in responding to computer security-related incidents; also called a *Computer Incident Response Team* (CIRT) or a CIRC (*Computer Incident Response Center, Computer Incident Response Capability*). This is a team providing services to a defined constituency. There are several acronyms used to describe teams providing similar types of services (e.g., CSIRC, CSRC, CIRC, CIRT, IHT, IRC, IRT, SERT, and SIRT). The generic term "CSIRT" has been widely adopted in the computer security community. SOURCE: SP 800-61

Computer Telephony Integration (CTI): A link between telephone systems and computers to facilitate incoming and outgoing calls; a physical link between a telephone and server.

computer virus attack: A program "infects" building control systems in much the same way as a biological virus infects humans. The typical virus makes copies of itself and inserts them into the code of other programs.

computer vision: A subfield of artificial intelligence (AI) focused on image processing. The goal is to achieve human-level understanding of an image.

Computerized Maintenance Management System (CMMS): A computerized database of information about a building's maintenance operations and includes the design, construction, commissioning, operations, maintenance, and replacement of equipment. The CMMS contains current and historical repair information for every piece of equipment in the building. This information is intended to help maintenance workers do their jobs more effectively (for example, determining which machines require maintenance and which storerooms contain the spare parts they need) and to help management make informed decisions (for example, calculating the cost of machine breakdown repair versus preventive maintenance for each machine, possibly leading to better allocation of resources).

Following is a list of the type of information stored in a CMMS that a hacker would want to access:

- **Equipment**: Nameplate, manufacturer, model number, capacity, condition, optimal ranges of operation, alarms for out-of-range conditions, criticality, spare parts on hand, and backup information.

- **Personnel**: Names, job title, e-mail address, phone number, certification and license, security clearance, permissions, passwords, labor rates, work schedule, and vacation schedule.

- **Maintenance operations**: Planned outages, equipment breakdowns, planned repairs, expansion plans, operating hours, and utility consumption.

- **Building tenants**: Complaints, including who complained, the name, job title, e-mail address, and phone number.

CMMS cyber-attack: Unfortunately, CMMS depends heavily on connectivity to the Internet as well as wireless communications to work efficiently. Building maintenance personnel are notified by the CMMS when equipment needs attention such as when a pump or valve malfunctions by generating and sending a work order to a mobile device. Personnel can access information wirelessly such as past maintenance history, preventive maintenance performed, all the specifications for the device including capacity, normal operating parameters and even whether spare parts are on hand and where they are located in the storage room. Some CMMS databases include tenant information such as who requested maintenance, the room number, and telephone number. Some databases contain information such as security clearance for staff, labor rates, vacation schedule and contact information. The CMMS would be a great tool to target maintenance personnel for spear phishing attacks.

When a hacker breaks into the CMMS, he can see a great deal of information about the building and how it is operated. A hacker can see which pieces of equipment are high-priority assets, which can be considered safety hazards and the trigger points for failure alarms and automatic shutdown. A hacker can see whether spare parts are on hand so he can target equipment that would take longer to repair. Another thing to consider is that the CMMS is typically tied directly to the BCS network making the CMMS a possible attack vector for hackers.

confidentiality: Preserving authorized restrictions on information access and disclosure, including means for protecting personal privacy and proprietary information. SOURCE: SP 800-53; SP 800-53A; SP 800-18; SP 800-27; SP 800-60; SP 800-37; FIPS 200; FIPS 199; 44 U.S.C., Sec. 3542

configuration control: Process of controlling modifications to hardware, firmware, software, and documentation to protect the information system against improper modification prior to, during, and after system implementation. SOURCE: CNSSI-4009; SP 800-37; SP 800-53

configuration management (CM): The management of changes made to a building control system's hardware, software, firmware, documentation, tests, test fixtures, and test documentation throughout the operational life of the building.

configuration property: Controller parameter used by the application that is usually set during installation and seldom changed.

congestion collapse attack: A condition that a packet-switched computer network can reach, when little or no useful communication is happening due to congestion. Generally, occurs at "choke points" in the network, where the total incoming traffic to a node exceeds the outgoing bandwidth.

consortium agreement: An agreement made by organizations to share processing and/or office facilities, if one member of the group suffers a cyber-attack.

contamination: Type of incident involving the introduction of data of one security classification or security category into data of a lower security classification or different security category. SOURCE: CNSSI-4009

content filtering: The process of monitoring communications such as e-mail and web pages, analyzing them for suspicious content, and preventing the delivery of suspicious content to users. SOURCE: SP 800-114

contingency plan: A plan for emergency response, back-up operations, and post- cyber-attack recovery for building control systems and installations. The contingency plan ensures minimal impact upon building operations in the event the building control system or facility is damaged or destroyed.

contingency planning: A plan that addresses how to keep an organization's building functions operating in the event of a cyber-physical attack.

control logic diagram: A graphic representation of control logic for the processes that make up the building control system.

control loop: A combination of field devices and control functions arranged so that a control variable is compared to a set point and returns to the process in the form of a manipulated variable. SOURCE: SP 800-82

control network: The networks connected to building equipment that control physical processes. A control network is subdivided into zones, and there can be multiple separate control networks in one building.

control server: A computer server that hosts the supervisory control system, typically a commercially available BCS or SCADA application.

controlled interface: A boundary with a set of mechanisms that enforces the security policies and controls the flow of information between interconnected information systems. SOURCE: CNSSI-4009; SP 800-37

controlled variable: A setpoint that the building control system attempts to keep at the predetermined value. The set point can be constant or variable.

controller: A device that monitors and controls the operation of a building control system. The operational conditions are output variables of the building control system, which are affected by adjusting input variables. The unit that manages information flow between data storage disks and a computer.

Controller Area Network (CAN bus): A automobile standard designed to allow microcontrollers and devices to communicate.

controller module: Communicates with a variety of devices, including Panel Bus I/O Modules and/or LonWorks devices (e.g., room controllers). Data points (of all types; e.g., internal virtual data points and hardware data points) are permitted. Typically, HVAC applications require an equal number of internal virtual data points and hardware data points. Controller Module and Panel Bus I/O Modules can typically be separated by up to 40 meters and can be powered by one or more external transformers. In case of a power failure, a capacitor typically saves RAM content and real-time clock up to 72 hours. SOURCE: Honeywell

Controlnet: Open network protocol for industrial automation applications.

cooling tower hack: A cooling tower is a building's heat rejection device that rejects waste heat to the atmosphere through the cooling of a water stream to a lower temperature. Cooling towers use the evaporation of water to remove process heat and cool the working fluid to near the wet-bulb air temperature or, in the case of closed circuit dry cooling towers, rely solely on air to cool the working fluid to near the dry-bulb air temperature. A basin heater in the cooling tower prevents the water from freezing during winter so cutting off power to the cooling tower heater could cause the water to freeze. That would prevent the unit from operating and could cause permanent damage.

cooling tower profile report: A cooling tower profile for each cooling tower, including:

- Total daily and monthly on-time (each fan).

- Number of on and off transitions (each fan).

- Maximum and minimum daily condenser water temperature and the time of occurrence for the current and previous months.

- Total daily and monthly makeup water consumption.

cookie: Data exchanged between an HTTP server and a browser (a client of the server) to store state information on the client side and retrieve it later for server use. SOURCE: CNSSI-4009

Continuity of Operations (COOP): A predetermined set of instructions or procedures that describe how an organization's mission-essential functions will be sustained within 12 hours and for up to 30 days as a result of a disaster event before returning to normal operations. SOURCE: SP 800-34

cooperative hot sites: A hot site owned by a group of organizations available to a group member should a cyber-attack occur.

corruption: A threat action that undesirably alters building control system operation by adversely modifying building control system functions or data.

cost benefit analysis: An assessment of the cost of providing data protection for a building control system versus the cost of compromising a building.

cost of ownership: The installation cost of equipment plus the cost of operating the equipment over its projected life span.

Counter-electronics High-powered Microwave Advanced Missile Project (CHAMP): A missile that can shut down computer systems and other nearby electronic systems from the sky through targeted emission of microwaves.

countermeasure: Any action, device, procedure, technique, or other measure that reduces the vulnerability of, or threat to a building control system. Reactive methods used to prevent an exploit from successfully occurring once a threat has been detected. Intrusion Prevention Systems (IPS) commonly employ countermeasures to prevent intruders form gaining further access to a computer network. Other countermeasures are *patches, access control lists* and *malware filters*. SOURCE: CNSSI-4009

countermeasures: That form of military science that, by the employment of devices and/or techniques, has as its objective the impairment of the operational effectiveness of undesirable or adversarial activity, or the prevention of espionage, sabotage, theft, or unauthorized access to or use of sensitive or classified information or information systems.

- **Defensive countermeasures** include actions to identify the source of hostile cyber activities protection/mitigation at the boundary (e.g., intrusion protection systems, preemptive blocks, blacklisting); hunting within networks (actively searching for insiders and other adversaries or malware); passive and active intelligence (including law enforcement) employed to detect cyber threats; and/or actions to temporarily isolate a system engaged in hostile cyber activities.

- **Offensive countermeasures** might include electronic jamming or other negation measures intended to disrupt an adversary's cyber capabilities during employment.

cover-coding: A technique to reduce the risks of eavesdropping by obscuring the information that is transmitted. SOURCE: SP 800-98

covert channel analysis: Determination of the extent to which the security policy model and subsequent lower-level program descriptions may allow unauthorized access to information. SOURCE: CNSSI-4009

covert channel attack: An unauthorized communication path that manipulates a communications medium in an unexpected, unconventional, or unforeseen way in order to transmit information without detection by anyone other than the entities operating the covert channel. SOURCE: CNSSI-4009

covert testing: Testing performed using covert methods and without the knowledge of the organization's IT staff, but with the full knowledge and permission of upper management. SOURCE: SP 800-115

covert timing channel: Covert channel in which one process signals information to another process by modulating its own use of system resources (e.g., central processing unit time) in such a way that this manipulation affects the real response time observed by the second process. SOURCE: CNSSI-4009

cracker: A hacker that breaks into computers for criminal gain. Also, someone who breaks into someone else's computer system, bypasses passwords or licenses in computer programs; or in other ways intentionally breaches computer security. Cracker also refers to those that reverse engineer software and modify it for grins and giggles.

cracking: Illegally breaking into a computer system.

crate & ship: Providing alternate processing capability during a cyber-attack, via contractual arrangements with an equipment supplier to ship replacement hardware within a specified time. Similar Terms: *guaranteed replacement, quick ship.*

credit card skimming attack: Hackers can use a device to steal credit card information in a legitimate credit card transaction. Credit card skimming devices are sometimes placed on ATMs or held in the hands of waiters or store employees. When a credit card is run through a skimmer, the device captures the credit card information.

crimeware: Malicious software designed to carry out illegal online activity. A class of malware designed to automate cybercrime.

crisis: A cyber-attack that dramatically impacts a building controls system's ability to operate properly.

crisis management: The coordination of an organization's response to a cyber-attack, in an effective, timely manner, with the goal of avoiding or minimizing damage to the building controls system's ability to operate.

crisis simulation: The process of testing an organization's ability to respond to a cyber-physical attack in a coordinated, timely, and effective manner, by simulating the occurrence of a specific cyber-attack.

critical functions: Building activities or operations that cannot be interrupted or made unavailable for an extended period of time without significantly jeopardizing the mission of the organization.

critical infrastructure: System and assets, whether physical or virtual, so vital to the United States that the incapacity or destruction of such systems and assets would have a debilitating impact on security, national economic security, national public health or safety, or any combination of those matters. SOURCE: Critical Infrastructures Protection Act of 2001, 42 U.S.C. 5195c(e) and CNSSI-4009

critical records: Records or data, which, if destroyed would cause considerable inconvenience and require replacement at considerable cost.

criticality level: Refers to the (consequences of) incorrect behavior of a system. The more serious the expected direct and indirect effects of incorrect behavior, the higher the criticality level. SOURCE: CNSSI-4009

cron: A Unix application that runs jobs at scheduled times of the day.

cross-domain capabilities: The set of functions that enable the transfer of information between security domains in accordance with the policies of the security domains involved. SOURCE: CNSSI-4009

cross-domain solution (CDS): A form of controlled interface that provides the ability to manually and/or automatically access and/or transfer information between different security domains. SOURCE: CNSSI-4009; SP 800-37

crossover cable: A cable that reverses the pairs of cables at the other end and can be used to connect devices directly together.

cross-site scripting (XSS) attack: A type of computer security vulnerability typically found in web applications. XSS vulnerabilities enable attackers to inject client-side script (typically Java) into web pages viewed by other users. A cross-site scripting vulnerability may be used by attackers to bypass access controls such as the same-origin policy. The effect may range from a petty nuisance to a significant security risk, depending on the sensitivity of the data handled by the vulnerable site and the nature of any security mitigation implemented by the site's owner.

cross-site request forgery (CSRF): A type of malicious exploit of a web site where unauthorized commands are transmitted from a user that the web site trusts. Unlike *cross-site scripting* (XSS), which exploits the trust a user has for a particular site, CSRF exploits the trust that a site has in a user's browser. Also known as *one-click attack* or *session riding*. For example, customers of a bank in Mexico were attacked in early 2008 with an image tag in e-mail. The link in the image tag changed the DNS entry for the bank in their ADSL router to point to a malicious web site impersonating the bank. SOURCE: Wikipedia

crowbar circuit: An electrical circuit used to prevent an overvoltage condition of a power supply from damaging the circuits attached to the power supply. It operates by putting a short circuit or low resistance path across the voltage output much as if one dropped a tool of the same name across the output terminals of the power supply. An active crowbar circuit can remove the short circuit when the transient is over allowing the device to resume normal operation.

CryptoDefense malware attack: One type of ransomware. Paying a ransom does not guarantee a victim will be able to access the data again and in many cases this doesn't happen.

CryptoLocker Trojan attack: A ransomware program distributed by e-mail attachments.

current sensing relay command switch: A non-intrusive device that detects current flowing through a cable or wire. Used for monitoring on and off status or proof of operation for motors driving fans and blowers, pumps, heating coils, and lighting.

current switch hack: Fixed trip point split-core and solid-core go/no go current switches are used for monitoring status of critical motors and electrical loads such as direct-drive units, exhaust fans, and other fixed loads and verifying lighting run times. These devices can be hacked if an attacker gains access to a BCS. SOURCE: Honeywell

cryptographic algorithm or hash: A well-defined computational procedure that takes variable inputs, including a cryptographic key, and produces an output. SOURCE: SP 800-21; CNSSI-4009

cryptography: The discipline that embodies the principles, means, and methods for the transformation of data in order to hide their semantic content, prevent their unauthorized use, or prevent their undetected modification. SOURCE: SP 800-59

cut-through: Switching where only the header of a packet is read before it is forwarded.

cyber-activism: Using the Internet to create, operate, and manage social activism. Also known as *hacktivism, Internet activism, cyber-dissidence, digital activism, online activism, online organizing, electronic advocacy,* and *e-activism.*

cyber-attack: An attack, via cyberspace, targeting an enterprise's use of cyberspace for the purpose of disrupting, disabling, destroying, or maliciously controlling a computing environment/infrastructure; or destroying the integrity of the data or stealing controlled information. SOURCE: CNSSI-4009 An ***active cyber-attack*** initiated from a computer against a BCS, ICS or SCADA system or individual computer that compromises the integrity or availability of the building control system, information stored on it, or the equipment controlled by it. ***Passive cyber-attacks*** are primarily reconnaissance to map a BAS, ICS or SCADA system, search for vulnerabilities or eaves dropping. An attack to commit a Cyber-Crime is considered a Cyber-Attack. Cyber-attacks are broken down into two categories:

- **Syntactic attacks** are straight forward; it is considered malicious software, which includes viruses, worms, and Trojan horses.

- **Semantic attack** is the modification and dissemination of correct and incorrect information. Information modified could have been done without the use of computers even though new opportunities can be found by using them. To set someone into the wrong direction or to cover your tracks through dissemination of incorrect information.

cyber-attack (untargeted & targeted): In ***untargeted cyber-attacks***, attackers indiscriminately target as many devices, services or users as possible. They do not care about who the victim is as there will be a number of machines or services with vulnerabilities. In a ***targeted cyber-attack***, your organization is singled out because the attacker has a specific interest in your business, or has been paid to target your building control system. The groundwork for the cyber-attack could take months so that a hacker can find the best route to deliver the exploit directly to your building control systems (or users). A targeted cyber-physical attack is often more damaging than an untargeted one because it has been specifically tailored to attack your building control systems, processes or personnel, in the office and sometimes at home. SOURCE: CERT: UK

cyber attack tree: A conceptual diagram showing how a computer system might be attacked by describing the threats and possible cyber-attacks to realize those threats. Cyber-attack trees lend themselves to defining an information assurance strategy and are increasingly being applied to industrial control systems and the electric power grid. Executing a strategy changes the cyber-attack tree. There are at least 150 ways to attack a BCS.

cyber booby trap: When a hacker embeds malware that is triggered by actions of the building maintenance staff. For example, the initial indication of a cyber-physical attack may be that the hacker turned off the water to a boiler. The maintenance personnel in the control room are unaware that the malware pumped all the water out of the boiler and turned up the heat. Once the boiler is superheated, the action of turning on the water triggers an explosion. The hacker needed the triggering action by building maintenance personnel to maximize the damage. Before turning the water back on, maintenance personnel should make sure that the power to the boiler is not turned on by manually checking the boiler and *not* rely on the BCS that has probably been rendered unreliable by the hacker.

cybercasting: The process by which a criminal can anonymously monitor a potential victim by watching as they sequentially upload valuable data about their possessions and their current geographical location (*geotag*).

cyber campaign: Denotes the time during which a given cyber force conducts a series of planned and coordinated cyber-attacks or other cyber operations in a given network environment (sometimes referred to as the *NEO*, or *network environment of operation*). A cyber campaign may be executed by a single actor or as a combined effort of multiple actors. A cyber campaign is a series of related cyber operations aimed toward a single, specific, strategic objective or result. A cyber campaign may take place over the course of just a few days or weeks, or it can last several months or even years.

cyber-collection: Refers to the use of cyber-warfare techniques in order to conduct espionage. Cyber-collection activities typically rely on the insertion of malware into a targeted network or computer in order to scan for, collect, and infiltrate sensitive information. Cyber-collection started as far back as 1996, when widespread deployment of Internet connectivity to government and corporate systems gained momentum. Since that time, there have been numerous cases of such activity. In addition to the state-sponsored examples, cyber-collection has also been used by organized crime for identity and e-banking theft and by corporate spies.

Common functionality of cyber-collection systems include the following:

- **Data scan:** local and network storage are scanned to find and copy files of interest, these are often documents, spreadsheets, design files such as AutoCAD files and system files such as the password file.

- **Capture location:** GPS, Wi-Fi, network information and other attached sensors are used to determine the location and movement of the infiltrated device.

- **Bug:** the device microphone can be activated in order to record audio. Likewise, audio streams intended for the local speakers can be intercepted at the device level and recorded.

- **Hidden Private Networks** that bypass the corporate network security. A computer that is being spied upon can be plugged into a legitimate corporate network that is heavy monitored for malware activity and at same time belongs to a private Wi-Fi network outside of the company network that is leaking confidential information off of an employee's computer. A computer like this is easily set up by a double-agent working in the IT department by install a second Wireless card in a computer and special software to remotely monitor an employee's computer through this second interface card without them being aware of a side-band communication channel pulling information off of his computer.

- **Camera:** the device cameras can be activated in order to covertly capture images or video.

- **Keylogger** and **mouse logger:** the malware agent can capture each keystroke, mouse movement, and click that the target user makes. Combined with screen grabs, this can be used to obtain passwords that are entered using a virtual on-screen keyboard.

- **Screen Grabber:** the malware agent can take periodic screen capture images. In addition to showing sensitive information that may not be stored on the machine, such as e-banking balances and encrypted web mail, these can be used in combination with the key and mouse logger data to determine access credentials for other Internet resources.

- **Encryption**: Collected data is usually encrypted at the time of capture and may be transmitted live or stored for later exfiltration. Likewise, it is common practice for each specific operation to use specific encryption and polymorphic capabilities of the cyber-collection agent in order to ensure that detection in one location will not compromise others.

- **Bypass encryption**: Because the malware agent operates on the target system with all the access and rights of the user account of the target or system administrator, encryption is bypassed. For example, interception of audio using the microphone and audio output devices enables the malware to capture to both sides of an encrypted Skype call.

- **Exfiltration**: Cyber-collection agents usually exfiltrate the captured data in a discrete manner, often waiting for high web traffic and disguising the transmission as secure web browsing. USB flash drives have been used to exfiltrate information from air gap protected systems. Exfiltration systems often involve the use of reverse proxy systems that anonymize the receiver of the data.

- **Replicate**: Agents may replicate themselves onto other media or systems; for example, an agent may infect files on a writable network share or install themselves onto USB drives in order to infect computers protected by an air gap or otherwise not on the same network.

- **Manipulate files and file maintenance**: Malware can be used to erase traces of itself from log files. It can also download and install modules or updates as well as data files. This function may also be used to place "evidence" on the target system; for example, to insert child pornography onto the computer of a politician or to manipulate votes on an electronic vote counting machine.

- **combination rules**: Some agents are very complex and are able to combine the preceding features to provide very targeted intelligence collection capabilities. For example, the use of GPS bounding boxes and microphone activity can be used to turn a smartphone into a *smart bug* that intercepts conversations only within the office of a target.

- **compromised cellphones**: Since, modern cellphones are increasingly similar to general purpose computer, these cellphones are vulnerable to the same cyber-collect attacks as computer systems, and are vulnerable to leak extremely sensitive conversational and location information to an attacker. Leaking of cellphone GPS location and conversational information to an attacker has been reported in a number of recent cyber stalking cases where the attacker was able to use the victim's GPS location to call nearby businesses and police authorities to make false allegations against the victim depending on his location. This can range from telling the restaurant staff information to tease the victim, or making false witness against the victim. For instance, if the victim were parked in a large parking lot the attackers may call and state that they saw drug or violence activity going on with a description of the victim and directions to their GPS location. Of course, plugging a cell phone into a computer to recharge the battery can infect the computer.

cyber-crime: Any crime to which a computer or computer technology has been used. The computer may be the target or it may have been used in the commission of a crime. There are two types of cyber-crime:

- **Type 1** cyber-crime is *a single event*. Such as when a user downloads a Trojan horse that installs a keystroke logger on his machine.

- **Type 2** cyber-crime is *an ongoing series of events*, involving repeated interactions with a target such as when a user is contacted in a chat room by someone who attempts to establish a relationship to commit a crime.

Use of any of the following to carry out a crime classifies it as a cyber-crime:

- Spamming

- Stalking, extortion, blackmail, or bullying

- Phishing

- Hacking

- Malware

- Exploiting a computer or building control system network vulnerabilities

- Social engineering and identity theft (fake e-mails, fake phone conversions using data obtained from the Internet, to get more information about a user such as banking information, credit card numbers, etc.)

cyber-drone: A cyber-drone can carry lightweight but powerful hacking platforms like Wi-Fi Pineapple and Raspberry Pi, packaged with an external battery pack and cellular connection, for powerful eavesdropping and man-in-the-middle attacks. A cyber-drone (or a swarm) can search for Wi-Fi wireless networks connected to a BCS at a facility and hack into networks when they are found. A cyber-drone can fly outside a skyscraper and find vulnerable networks with minimal interference. For example, a cyber-drone can land on a roof and target wireless printers because they often are the weak link in a company's wireless network. Wireless printers are typically supplied with the Wi-Fi connection open by default, and many companies forget to close this hole when they add the device to their Wi-Fi networks. This open connection potentially provides an access point for outsiders to connect to a BCS network. It is possible for a cyber-drone to shut down computer systems and other nearby electronic systems from the sky through targeted emission of microwaves.

Anti-drone technology is beginning to come on the market for WLAN customers as cyber-drones become attached to more verified network attacks. Fluke Networks has released the first *cyber-drone detection signature* as an update to its AirMagnet Enterprise wireless IDS/IPS product that alerts customers to drone-specific signals. Cyber-drones are controlled via an ad hoc network and AirMagnet can detect the command-and-control signaling. The AirMagnet also can detect video transmission streams. Once alerted, the network administrator can either attempt to locate the drone and its operator, or take RF or WLAN system-level countermeasures.

cyber event: An observable occurrence that raises the suspicion that a cyber-incident may be occurring. Cyber events fall into four categories:

- **True Positive**: Something bad happened and the NIDS caught it.

- **True Negative**: The event is benign and no alert was generated.

- **False Positive**: The NIDS alert sounded, but the event was not malicious.

- **False Negative**: Something bad happened, but the NIDS did not catch it.

cyber-exercise: A planned event during which an organization simulates a cyber-attack to test capabilities such as preventing, detecting, mitigating, responding to or recovering from the cyber-attack.

cyber-hygiene: Steps you take to improve your cybersecurity and better protect yourself online.

cyber-incident: Actions taken through the use of computer networks that result in an actual or potentially adverse effect on an information system and/or the information residing therein. The difference between a *cyber-incident* and a *cyber-event* is that you are obligated to report a cyber-incident because it is now a law enforcement issue. A cyber-event merely raises the suspicion that an incident may be occurring. SOURCE: CNSSI-4009

cyber-infiltration: There are several common ways to infect or access the target:

- **An injection proxy** is a system that is placed upstream from the target individual or company, usually at the Internet Service Provider, that injects malware into the targets system. For example, an innocent download made by the user can be injected with the malware executable on the fly so that the target system then is accessible to hackers.

- **spear phishing**: A carefully crafted e-mail is sent to the target in order to entice them to install the malware via a Trojan document or a drive by attack hosted on a web server compromised or controlled by the malware owner.

- **Surreptitious entry** may be used to infect a system. In other words, the spies carefully break into the target's office and install the malware on the target's system.

- **An upstream monitor** or sniffer is a device that can intercept and view the data transmitted by a target system. Usually this device is placed at the Internet Service Provider. The Carnivore system is a famous example of this type of system. Based on the same logic as a telephone intercept, this type of system is of limited use today due to the widespread use of encryption during data transmission.

- **A wireless infiltration** system can be used in proximity of the target when the target is using wireless technology. This is usually a laptop-based system that impersonates a Wi-Fi or 3G base station to capture the target systems and relay requests upstream to the Internet. Once the target systems are on the network, the system then functions as an injection proxy or as an upstream monitor in order to infiltrate or monitor the target system.

- A **USB** Key preloaded with the malware infector may be given to or dropped at the target site.

cyber-infrastructure: Includes electronic information and communications systems and services and the information contained in these systems and services. Information and communications systems and services are composed of all hardware and software that process, store, and communicate information, or any combination of all of these elements. Processing includes the creation, access, modification, and destruction of information. Storage includes paper, magnetic, electronic, and all other media types. Communications include sharing and distribution of information. For example, computer systems; control systems (e.g., supervisory control and data acquisition—SCADA); networks, such as the Internet; and cyber services (e.g., managed security services) are part of cyber-infrastructure. SOURCE: NISTIR 7628

cyber liability insurance: Commercial insurance policy providing general coverage.

Cyber Operational Preparation of the Environment (C-OPE): Non-intelligence enabling functions within cyberspace conducted to plan and prepare for potential follow-on military operations. C-OPE includes but is not limited to identifying data, system/network configurations, or physical structures connected to or associated with the network or system (to include software, ports, and assigned network address ranges or other identifiers) for the purpose of determining system vulnerabilities; and actions taken to assure future access and/or control of the system, network, or data during anticipated hostilities.

cyber operations: Use of offensive and defensive activities for achieving objectives in cyberspace.

cyber-physical attack: A classic cyber-physical attack would be when a hacker is able to damage building equipment by sending destructive commands over the BCS to the equipment that change the setpoints above dangerous levels for which the equipment has not been designed, such as too high pressure or dangerously high temperature. A cyber-physical attack is an attack that does actual physical damage to vulnerable physical systems.

cyber-physical attack engineering: Designing an attack scenario to exploit a particular physical process requires a solid engineering background and in-depth *destructive* knowledge of the target SCADA system. Hacking a chemical plant, for example requires knowledge of physics, chemistry and engineering, as well as a great deal about how the network is laid out, and a keen understanding of process-aware defensive systems. This represents a high (but not insurmountable) barrier to entry to garden-variety script kiddies, but is not a major obstacle for a foreign intelligence service.

cyber-physical recovery procedures: What a facility manager should do when he suspects a cyber-physical attack on his building controls system, how to detect and confirm a cyber-physical attack is underway, what he needs to do (and not do) to mitigate damage to the facility, and how to recover from the cyber-physical attack. My other book, *Cyber-Physical Attack Recovery Procedures* (Apress, 2016), covers this subject.

cyber-physical systems (CPS): Engineered systems that are built from, and depend upon, the seamless integration of computational algorithms and physical components. Traditional analysis tools are unable to cope with the full complexity of CPS or adequately predict system behavior. As the Internet of Things (IoT) scales to billions of connected devices: with the capacity to sense, control, and otherwise interact with the human and physical world: the requirements for dependability, security, safety, and privacy grow immensely. One barrier to progress is the lack of appropriate science and technology to conceptualize and design for the deep interdependencies among engineered systems and the natural world. SOURCE: National Science Foundation

cyber response plan: Provides guidance to prepare for and respond to a cyber incident regardless of source. The document also suggests ways to learn from incidents and to strengthen the system against potential attacks. Provides written instructions on assembling an Incidence Response Team, assigning duties and responsibilities and method for documenting the nature and scope of a cyber-physical attack.

Cyber Resilience Review (CRR): DHS software tool used to evaluate operational resilience and cyber-security practices and to assess risk.

cyberspace: A global domain within the information environment consisting of the interdependent network of information systems infrastructures including the Internet, telecommunications networks, computer systems, and embedded processors and controllers. SOURCE: CNSSI-4009

cybersecurity: The ability to protect or defend the use of cyberspace from cyber-attacks. SOURCE: CNSSI-4009

cybersecurity budget: If this isn't a line in your *facility budget*, make it so (1% of facility O&M budget).

cyber security evaluation tool (CSET): DHS software tool that allows users to assess their controls and network security practices against industry standards and makes recommendations.

cyberspace superiority: The degree of dominance in cyberspace by one force that permits the secure, reliable conduct of operations of that force, and its related land, air, sea, and space forces at a given time and sphere of operations without prohibitive interference by an adversary.

cyber spying: Cyber espionage is the act or practice of obtaining secrets without the permission of the holder of the information (personal, sensitive, proprietary, or of classified nature), from individuals, competitors, rivals, groups, governments, and enemies for personal, economic, political, or military advantage using methods on the Internet, networks, or individual computers through the use of cracking techniques and malicious software including Trojan horses and spyware. It may wholly be perpetrated online from computer desks of professionals on bases in other countries or may involve infiltration at home by computer trained conventional spies and moles or in other cases may be the criminal handiwork of amateur malicious hackers and software programmers.

cyberterrorism: The use of computer network tools to shut down critical national infrastructures (such as energy, transportation, government operations) or to coerce or intimidate a government or civilian population. The end result of both cyberwarfare and cyberterrorism is the same, to damage critical infrastructures and building control systems linked together within the confines of cyberspace.

cyber tools:

- **commodity capability**: Cyber tools and techniques openly available on the Internet (off-the-shelf) that are relatively simple to use. This includes tools designed for security specialists (such as building control system penetration testers) that can also be used by attackers as they are specifically designed to scan for publicly known vulnerabilities in operating systems and applications. Poison Ivy is a good example of a commodity tool; it is a readily available Remote Access Tool (RAT) that has been widely used for a number of years.

- **bespoke capability**: Cyber tools and techniques that are developed and used for specific purposes, and thus require more specialist knowledge. This could include malicious code ('exploits') that take advantage of software vulnerabilities (or bugs) that are not yet known to vendors or anti-malware companies, often known as *zero-day* exploits. It could also include undocumented software features, or poorly designed applications. Bespoke capabilities usually become commodity capabilities once their use has been discovered, sometimes within a few days. By their very nature, the availability of bespoke tools is not advertised as once released they become a commodity. SOURCE: CERT-UK

cyberwarfare: Actions by a nation-state to penetrate another nation's computers or networks for the purposes of causing damage or disruption. The fifth domain of warfare (the others are: *land, sea, air,* and *space*).

Cyborg Unplug: A plug-and-play network appliance that automatically detects and disconnects a range of Internet-connected surveillance devices including Dropcam, Google Glass, and Wi-Fi-enabled drones by breaking uploads and streams. It sniffs the air for wireless signatures from devices known to pose a risk to personal privacy, Cyborg Unplug will disconnect them, stopping them from streaming video, audio, and data to the Internet and it sends an e-mail alert. Most wireless devices used for surveillance; stream data to a machine on the Internet or in a nearby room allowing for remote surveillance while ensuring the offending device contains no evidence (files) of the abuse.

cycle of concentration: The degree to which dissolved solids and other impurities build up in recirculating water in a cooling tower. Specifically, cycles of concentration equal the ratio of the concentration of the dissolved solids (chlorine, sulfates) in the recirculating water to the concentration of the same material in the make-up water.

cycle time: The time, usually expressed in seconds, for a controller to complete one control loop where sensor signals are read into memory, control algorithms are executed, and corresponding control signals are transmitted to actuators that create changes in the process resulting in new sensor signals.

cycle timer hack: A timing device that can be preset to turn off and on at specific intervals. Hack this and an attacker can cause things to turn off when they should be on and vice versa.

cyclic redundancy check (CRC): An error-detecting code commonly used in digital networks and storage devices to detect accidental changes to raw data. Blocks of data entering these systems get a short check value attached, based on the remainder of a polynomial division of their contents. On retrieval, the calculation is repeated and corrective action can be taken against presumed data corruption if the check values do not match.

CHAPTER 5

D

daemon: A program that runs as a background process is often started at the time the building control system boots and runs continuously without intervention from any of the users on the building control system. The daemon program forwards the requests to other programs (or processes) as appropriate. The term *daemon* is a Unix term, although many other operating systems provide support for daemons; however, they're sometimes called other names. Windows, for example, refers to daemons and system agents and services.

damage assessment: The process of assessing damage following a cyber-physical attack on BCS hardware, vital records, facilities, and so forth, and determining what can be salvaged and what must be replaced.

damper hack: A valve or plate that stops or regulates the flow of air inside a duct, chimney, VAV box, air handler, or other air handling equipment. A damper may be used to cut off central air conditioning (heating or cooling) to an unused room, or to regulate it for room-by-room temperature and climate control. Its operation can be manual or automatic. Manual dampers are turned by a handle on the outside of a duct. Automatic dampers are used to regulate airflow constantly and are operated by electric or pneumatic motors, in turn controlled by a thermostat or the building controls system. Automatic or motorized dampers may also be controlled by a solenoid, and the degree of air-flow calibrated, according to signals from the thermostat going to the actuator of the damper in order to modulate the flow of air-conditioned air. A hacker can use the BCS to close select dampers and throw the HVAC out of balance.

damper schedule: The damper schedule indicates each damper's unique identifier, type (opposed or parallel blade), nominal and actual sizes, orientation of axis and frame, direction of blade rotation, actuator size and spring ranges, operation rate, positive positioner range, location of actuators and damper end switches, arrangement of sections in multi-section dampers, and methods of connecting dampers, actuators, and linkages. The damper schedule indicates the AMCA 511 maximum leakage rate at the operating static-pressure differential.

data asset:

- Any entity that is comprised of data. For example, a database is a data asset that is comprised of data records. A data asset may be a system or application output file, database, document, or web page. A data asset also includes a service that may be provided to access data from an application. For example, a service that returns individual records from a database would be a data asset. Similarly, a web site that returns data in response to specific queries (e.g., www.weather.com) would be a data asset.

- An information-based resource.

SOURCE: CNSSI-4009

© Luis Ayala 2016
L. Ayala, *Cybersecurity Lexicon*, DOI 10.1007/978-1-4842-2068-9_5

database: A repository of digital information that holds building or plant-wide information including manufacturing process data, personnel data, and financial data.

data breach: The unauthorized disclosure of sensitive information to a party that is not authorized to have the information.

data center recovery: Disaster recovery that deals with the restoration, at an alternate location, of data center services.

data center relocation: The relocation of an organization's entire data center operation.

data custodian: The entity currently using or manipulating the data and taking responsibility for the data temporarily.

data diode: A network appliance or device allowing data to travel only in one direction, used in guaranteeing information security. They are most commonly found in high security environments such as defense, where they serve as connections between two or more networks of differing security classifications. Also referred to as a *unidirectional security gateway* or *unidirectional network*.

datagram: The message units that the Internet Protocol deals with and that the Internet transports. A datagram or packet needs to be self-contained without reliance on earlier exchanges because there is no connection of fixed duration between the two communicating points. The Datagram service is considered unreliable. This kind of protocol is known as *connectionless*.

data historian: A centralized database supporting data analysis using statistical process control (SPC).

data integrity: The property that data has not been altered in an unauthorized manner. Data integrity covers data in storage, during processing, and while in transit. SOURCE: SP 800-27

data loss attack: The result of unintentionally or accidentally deleting data, forgetting where it is stored, or exposure to an unauthorized party. SOURCE: NICCS

data mining: Technique used to analyze a vast amount of information.

Day Zero: The *Day Zero* or *Zero Day* is the day a new vulnerability is made known. In some cases, a "zero-day exploit" is referred to an exploit for which no patch is available yet. ("day one" > day at which the patch is made available).

DC servo drive hack: A type of drive that works specifically with servo motors. It transmits commands to the motor and receives feedback from the servo motor resolver or encoder. A hacker can cause the drive to transmit false commands to servo motors or ignore feedback from the motors.

deauthentication packet attack: This attack sends disassociating packets to one or more clients that are currently associated with a particular Wi-Fi access point thereby breaking the connection. The deauthentication packets are sent directly from a PC to the clients, so the attacker must be physically close enough to the clients for wireless transmissions to reach them.

decapsulation: Stripping off one layer header and passing the rest of the packet up to a higher layer on the stack.

decertification: Revocation of the certification of an information system item or equipment for cause. SOURCE: CNSSI-4009

declaration fee: A one-time fee, charged by an Alternate Facility provider, to a customer who declares a cyber-attack. Similar term: *notification fee*. Note: *Some recovery vendors apply the declaration fee against the first few days of recovery.*

decoy password: Some software programs such as Vault Apps use a decoy password so when someone sees the photo vault on your smartphone, you tell them the decoy password and it opens up to a set of fake secret photos so people lose their curiosity.

DecryptCryptoLocker: FireEye tool used to unlock ransomware victims' machines.

decryption: The process of changing ciphertext into plaintext using a cryptographic algorithm and key. SOURCE: SP 800-21

dedicated line: A pre-established point-to-point communication link between computer terminals, or between distributed processors that does not require dial-up access.

dedicated mode: Information systems security mode of operation wherein each user, with direct or indirect access to the system, its peripherals, remote terminals, or remote hosts, has all of the following:

- valid security clearance for all information within the system,

- formal access approval and signed nondisclosure agreements for all the information stored and/or processed (including all compartments, subcompartments, and/or special access programs), and

- valid need-to-know for all information contained within the information system. When in the dedicated security mode, a system is specifically and exclusively dedicated to and controlled for the processing of one particular type or classification of information, either for full-time operation or for a specified period of time.

SOURCE: CNSSI-4009

dedicated outdoor air system (DOAS) hack: A type of HVAC system that consists of two parallel systems: a dedicated outdoor air ventilation system that handles latent loads and a parallel system to handle sensible loads. Conventional HVAC systems, such as VAV systems serve multiple zones and have potential problems in terms of poor thermal comfort and possible microbial contamination. DOAS provides dedicated ventilation rather than ventilation as part of conditioned air. Hacking the DOAS would allow an attacker to cut off outside ventilation to the building.

deep packet inspection (DPI): A form of computer network packet filtering that examines the data part (and possibly also the header) of a packet as it passes an inspection point, searching for protocol non-compliance, viruses, spam, intrusions, or defined criteria to decide whether the packet may pass or if it needs to be routed to a different destination, or, for the purpose of collecting statistical information. Also called *complete packet inspection*.

defacement attack: The addition, removal, or change of content, in a deliberate attempt to damage or embarrass the web site owner.

defense-in-breadth: A planned, systematic set of multidisciplinary activities that seek to identify, manage, and reduce risk of exploitable vulnerabilities at every stage of the system, network, or subcomponent life cycle (system, network, or product design and development; manufacturing; packaging; assembly; system integration; distribution; operations; maintenance; and retirement). SOURCE: CNSSI-4009

defense-in-depth: Information security strategy integrating people, technology, and operations capabilities to establish variable barriers across multiple layers and dimensions of the organization. SOURCE: CNSSI-4009; SP 800-53

defensive counter-cyber (DDC): All defensive countermeasures designed to detect, identify, intercept, and destroy or negate harmful activities attempting to penetrate or attack through cyberspace. DCC missions are designed to preserve friendly network integrity, availability, and security, and protect friendly cyber capabilities from attack, intrusion, or other malicious activity by pro-actively seeking, intercepting, and neutralizing adversarial cyber means that present such threats. DCC operations may include military deception via honeypots and other operations; actions to adversely affect adversary and/or intermediary systems engaged in a hostile act/imminent hostile act; and redirection, deactivation, or removal of malware engaged in a hostile act/imminent hostile act.

delay-tolerant networking (DTN): An approach to computer network architecture that seeks to address the technical issues in heterogeneous network that may lack continuous network connectivity. Examples of such networks are those operating in mobile or extreme terrestrial environments, or planned networks in space. Recently, the term *disruption-tolerant networking* has gained usage in the United States due to support from DARPA, which has funded many DTN projects. Disruption may occur because of the limits of wireless radio range, sparsity of mobile nodes, energy resources, attack, and noise.

deleted file: A file that has been logically, but not necessarily physically, erased from the operating system, perhaps to eliminate potentially incriminating evidence. Deleting files does not always necessarily eliminate the possibility of recovering all or part of the original data. SOURCE: SP 800-72

Defense Security Service (DSS): Provides security oversight of the facilities performing on classified contract work for US federal agencies within the NISP. DSS oversees the protection of US and foreign classified information and technologies in the hands of cleared industry under the National Industrial Security Program by providing professional risk management services. As functional manager for the US Department of Defense (DoD), DSS provides security education, training, certification, and professional development for DoD and for other US government personnel, contractor employees, and representatives of foreign governments.

demand control ventilation (DCV): A control method used to automatically adjust ventilation equipment to increase or decrease the volume of fresh air taken into a building by air conditioning equipment.

demand response (DR): According to the Federal Energy Regulatory Commission, demand response is an "...action taken to reduce electricity demand in response to price, monetary incentives, or utility directives so as to maintain reliable electric service or avoid high electricity prices." In the real world, DR is a voluntary program that compensates customers for reducing their electricity use during periods of high demand or when reliability of the commercial power grid is threatened.

demand response automation server (DRAS): A DRAS processes demand response event information with predefined DR control actions, and commands DR control points to execute control actions such as turning off lights, raising thermostat settings, limiting fan speed, locking out chillers, and so forth. Maintenance personnel configure the DRAS server communication parameters, control points, and predefine control actions. A hacker can too.

demilitarized zone (DMZ): An interface on a routing firewall that is similar to the interfaces found on the firewall's protected side. Traffic moving between the DMZ and other interfaces on the protected side of the firewall still goes through the firewall and can have firewall protection policies applied. SOURCE: SP 800-41

denial-of-service (DoS) attack: An attack that prevents or impairs the authorized use of networks, systems, or applications by exhausting resources. SOURCE: SP 800-61

The following are some of the different forms of DoS attack:

- **Teardrop**: Sending irregularly shaped network data packets.

- **Buffer Overflow**: Flooding a server with an overwhelming amount data.

- **Smurf**: Tricking computers to reply to a fake request, causing much traffic.

- **Physical**: Disrupting a physical connection, such as a cable or power source.

▨ **Caution**   It is believed that a DDoS attack was used as a means of distracting a company's cyber-defense team while criminals went about stealing customer account information in the United Kingdom in October 2015.

---

device fingerprint: Information collected about a remote computing device for the purpose of identification. Device fingerprints can be used to fully or partially identify individual users or devices even when cookies are turned off. Also called a *machine fingerprint* or *browser fingerprint*.

device registration manager: The management role that is responsible for performing activities related to registering users that are devices. SOURCE: CNSSI-4009

Departmental Recovery Team: Staff responsible for performing recovery procedures specific to their department.

departure manager (DMAN): A system aid for the air traffic control at airports that calculates planned departure flow to maintain an optimal throughput at the runway, reduce queuing, and distribute information to various stakeholders at the airport.

device object: Every BACnet device requires one device object, whose properties represent the network visible properties of that device. Every device object requires a unique Object_Identifier number on the BACnet internetwork. This number is often referred to as the *device instance* or *device ID*.

DeviceNet: A network system used in the automation industry to interconnect control devices for data exchange.

diagnostic server attacks: An attacker can execute the following attacks without any authentication required while maintaining stealthiness such as remote memory dump, remote memory patch, remote calls to functions and remote task management.

diagnostics: Information concerning known failure modes and their characteristics. Such information can be used in troubleshooting and failure analysis to help pinpoint the cause of a failure and define suitable corrective action.

dial backup: The use of dial-up communication lines as a backup to dedicated lines.

dial-up line: A communication link between computer terminals that is established by dialing a specific telephone number.

dictionary attack: When a cyber-attack utilizes a dictionary to crack a password. Words from the dictionary are input in the password field to try to guess the password. Programs and tools allow hackers to easily try combinations of words in the dictionary to crack a user's password. We recommend personnel use passwords that do not contain simple words found in a dictionary.

diesel generator set hack: A packaged diesel engine, generator and various ancillary devices (such as base, canopy, sound attenuation, building control system, circuit breakers, jacket water heater and starter) is referred to as a *generating set* or a *genset*. Hacking the genset prevents the backup power from operating when it is needed most.

differential power analysis (DPA): An analysis of the variations of the electrical power consumption of a cryptographic module, using advanced statistical methods and/or other techniques, for the purpose of extracting information correlated to cryptographic keys used in a cryptographic algorithm. SOURCE: FIPS 140-2

digest authentication: Allows a web client to compute MD5 hashes to prove it has the password.

digital audio tape (DAT): A digital magnetic tape format originally developed for audio recording now used for computer backup tape.

digital certificate: An electronic "credit card" that establishes your credentials when doing business or other transactions on the Web. It is issued by a certification authority. It contains your name, a serial number, expiration dates, a copy of the certificate holder's public key (used for encrypting messages and digital signatures), and the digital signature of the certificate-issuing authority so that a recipient can verify that the certificate is real.

digital envelope: An encrypted message with the encrypted session key.

digital linear tape (DLT): A serpentine magnetic tape technology used for midrange to high-end tape backup requirements of networks and servers.

digital picture frame attack: Some digital picture frames manufactured in the Far East have been found to contain a virus that steals passwords and financial information, blocks security and replicates when loading pictures from a home computer using a USB drive.

digital protection and control device (DPCD): A hardware device that monitors for the rapid out-of-phase condition associated with an AUORA event between an electrical substation and its loads. The device isolates the substation from its loads before the torque of the grid can be applied to the equipment.

digital reputation: Reputation defined by behaviors online and by the content a person posts. Photos, blog posts, and social network interactions determine how a person will be perceived by others online.

digital rights management: A type of access control technology to protect and manage use of copyrighted digital content.

digital signature: An asymmetric key operation where the private key is used to digitally sign an electronic document and the public key is used to verify the signature. Digital signatures provide authentication and integrity protection. SOURCE: SP 800-63

digital signature algorithm (DSA): Asymmetric algorithms used for digitally signing data. SOURCE: SP 800-49

digital signature standard: The US Government standard that specifies the Digital Signature Algorithm (DSA) using asymmetric cryptography.

direct-access attack: A direct-access attack means gaining physical access to a building control system and performing various functions or installing various types of devices to compromise building operations. The attacker can install a virus, worm, or Trojan horse, download building operations data, survey building activity, or change the operating parameters of building equipment to the point of equipment failure.

Direct Digital Controls (DDC) hack: DDC is the automated control of a condition or process by a computer. All instrumentation is gathered by various analog and digital devices that use the network to transport these signals to the central controller. The central computer then follows all of its production rules and causes action requests to be sent via the same network to valves, actuators, and other HVAC components that can be adjusted. Hack the DDC and you control the processes completely.

disassembly: Taking a binary software program and deriving the source code from it.

disaster: Any event that creates an inability to provide critical building functions.

disaster prevention: Measures used to prevent, detect, or contain incidents that, if unchecked, could result in a cyber-physical attack on a BCS.

disaster prevention checklist: A questionnaire used to assess cyber-attack prevention measures in areas of operations such as overall security, software, data files, data entry reports, computers and personnel.

disaster recovery: The ability to respond to a cyber-physical attack on building services by implementing a cyber-physical attack recovery plan to restore critical building functions.

disaster recovery administrator: Individual responsible for documenting cyber-physical attack recovery activities and tracking building recovery progress.

disaster recovery coordinator: The disaster recovery coordinator may be responsible for overall recovery of critical building functions.

disaster recovery period: The time period between a cyber-physical attack and a return to normal building functions, during which the disaster recovery plan is employed.

disaster recovery plan (DRP): A written plan for recovering one or more information systems at an alternate facility in response to a major hardware or software failure or destruction of facilities. SOURCE: SP 800-34

disaster recovery planning: The technological aspect of building continuity planning. The advance planning and preparations that are necessary to minimize loss and ensure continuity of the critical building functions of an organization in the event of a cyber-physical attack.

disaster recovery software: An application program developed to assist an organization in writing a comprehensive cyber-physical attack recovery plan.

disaster recovery teams (building recovery teams): A group of teams ready to take control of the recovery operations if a cyber-physical attack should occur.

discretionary access control: Something the user can manage, such as a document's password.

disk array: Arrangement of two or more hard drives in a RAID or daisy-chain configuration to improve speed and protect against data against loss.

disk imaging: Generating a bit-for-bit copy of the original media, including free space and slack space. SOURCE: SP 800-86

disruption: An unplanned event that causes the general system or major application to be inoperable for an unacceptable length of time (e.g., minor or extended power outage, extended unavailable network, or equipment or facility damage or destruction). SOURCE: CNSSI-4009 An unplanned event that causes an information system to be inoperable for a length of time (e.g., minor or extended power outage, extended unavailable network, or equipment or facility damage or destruction). SOURCE: SP 800-34

distance vector: Measure the cost of routes to determine the best route to all known networks.

distributed computing environment: Middleware standards that define the method of communication between clients and servers in a cross-platform computing environment; enables a client program to initiate a request that can be processed by a program written in a different computer language and housed on a different computer platform.

distributed controls system (DCS): Distributed controls for an industrial process or plant, wherein control elements are distributed throughout the building control system. This is in contrast to non-distributed building control systems, which use a single controller at a central location. In a DCS, a hierarchy of controllers is connected by communications networks for command and monitoring.

distributed denial-of-service (DDoS) attack: A denial-of-service technique that uses numerous hosts to perform the attack. SOURCE: SP 800-61; CNSSI-4009 Sometimes called an *amplification attack*.

distributed energy resources (DER): Modular and flexible power systems that are located close to the load they serve, typically with capacities of 10 megawatts (MW) or less. They are typically low-voltage AC grids, often use diesel generators, and operate autonomously to provide grid resilience and mitigate grid disturbances. Becoming increasingly important because digital equipment requires extremely reliable electricity.

distributed generation (DG): Power generation at the point of consumption. Generating power on-site, typically using renewable energy sources, eliminates the dependence on the commercial power grid that is vulnerable to cyber physical attack.

distributed I/O: Eliminates expensive point-to-point wires by networking process signals onto one digital communication link.

distributed scans: Scans that use multiple source addresses to gather information.

disturbance: An undesired change in a variable being applied to a building control system designed to adversely affect the value of a controlled variable.

DNS forgery attack: A hacker with access to a network can easily forge responses to the computer's DNS requests.

DNS sinkhole: Also known as a *sinkhole server*, *Internet sinkhole*, or *BlackholeDNS*. A DNS server that gives out false information, to prevent the use of the domain names it represents. A *sinkhole* is a standard DNS server that has been configured to hand out non-routable addresses for all domains in the sinkhole, so that every computer that uses it will fail to get access to the real web site. The higher up the DNS server is, the more computers it will block. Some of the larger botnets have been made unusable by TLD sinkholes that span the entire Internet. DNS Sinkholes are effective at detecting and blocking malicious traffic, and used to combat bots and other unwanted traffic.

DNS spoofing: A computer hacking attack, whereby data is introduced into a Domain Name System (DNS) resolver's cache, causing the name server to return an incorrect IP address, diverting traffic to the attacker's computer (or any other computer). Also called *DNS cache poisoning*.

domain: A set of subjects, their information objects, and a common security policy. SOURCE: SP 800-27 An environment or context that includes a set of system resources and a set of system entities that have the right to access the resources as defined by a common security policy, security model, or security architecture. See Security Domain. SOURCE: CNSSI-4009; SP 800-53; SP 800-37

domain controller: A Microsoft Windows server that has Active Directory Domain Services installed which manages domain information, such as login identification and passwords.

domain hijacking attack: A cyber-attack that takes over a domain by first blocking access to the domain's DNS server and then putting the hacker's server in its place.

doorknob-rattling attack: A hacker attempts a very few common username and password combinations on several computers resulting in very few failed login attempts. This attack can go undetected unless the data related to login failures from all the hosts are collected and aggregated to check for doorknob-rattling from any remote destination.

doorway page: Web pages that are created for spamdexing. This is for spamming the index of a search engine by inserting results for particular phrases with the purpose of sending visitors to a different page. They are also known as *bridge pages*, *portal pages*, *jump pages*, *gateway pages*, *entry pages*, and other names. Doorway pages that redirect visitors without their knowledge use some form of cloaking.

Dorkbot: Dorkbot is a botnet used to steal online payment, participate in distributed denial-of-service (DDoS) attacks, and deliver other types of malware to victims' computers. Dorkbot is commonly spread via malicious links sent through social networks instant message programs or through infected USB devices. Dorkbot's backdoor functionality allows a remote attacker to exploit infected systems. A system infected with Dorkbot may be used to send spam, participate in DDoS attacks, or harvest users' credentials for online services, including banking services.

doxing: Publicly releasing a person's identifying information including full name, date of birth, address, and pictures typically retrieved from social networking site profiles. SOURCE: FBI Internet Social Networking Risks

draining: A form of urban exploration (urbex) when exploring drains. Also commonly referred to as *infiltration*, *urban spelunking*, *urban rock climbing*, *urban caving*, or *building hacking*. Sewers are the most dangerous locations to explore due to risk of poisoning by toxic gas such as methane and hydrogen sulfide.

drift eliminator: Cooling towers use drift eliminators to minimize drift mass airflow and prevent escape of entrained water droplets that might contain Legionella bacteria. A plenum is used to maintain airflow within tolerances of design throughput, particularly at the center of the eliminator bank in counterflow towers and at the upper portions of the eliminator bank in crossflow towers.

drive-by download attack: Malware installed on a target computer or other device as soon as a user visits a compromised web site.

droop speed control: Used as a countermeasure to cyber-attack when disconnecting the electrical power grid from the Internet. In electrical power generation, a speed control mode of a prime mover driving a synchronous generator connected to an electrical grid. This mode allows synchronous generators to run in parallel, so that loads are shared among generators in proportion to their power rating.

Dropcam: Dropcam and its successor Nest Cam allow people to view live video through a cloud-based service. Dropcam provides free live streaming video that is accessible through a web app and mobile apps for iOS and Android.

Dropper attack: Computer malware that allows attackers to open a backdoor to install another malware program to an infected machine to implement additional functionality. *Briba* is a good example.

duct leakage test: A diagnostic tool designed to measure the airtightness of forced air HVAC ductwork. A duct leakage tester consists of a calibrated fan for measuring an air flow rate and a pressure sensing device to measure the pressure created by the fan flow. The combination of pressure and fan flow measurements are used to determine the ductwork airtightness. SOURCE: Wikipedia

due care: The responsibility that manages and their organizations have a duty to provide for information security to ensure that the type of control, the cost of control, and the deployment of control are appropriate for the system being managed. SOURCE: SP 800-30 A legal concept distinct from due diligence. "Did the company know of the vulnerability and the damage it could cause (including injury)?"

due diligence: The requirement that an organization deploy protection to prevent a cyber-attack from occurring. A legal concept distinct from due care. "Did the company do anything to prevent the damage the vulnerability could cause (including injury)?"

DumpSec: A security tool that dumps a variety of information about a building control system's users, file system, registry, permissions, password policy, and services.

dumpster diving: Obtaining passwords and organization's directories by searching through discarded trash bin. Also referred to as *skipping*.

dust collector: Dust collection systems captures dust particles of all sizes out of the air stream in manufacturing facilities. A dust bin positioned underneath the funnel captures the dust and chips in a trash container. These are the primary pieces of equipment associated with dust explosions. It is estimated that 70% to 80% of all industrial explosions where combustible dusts are handled are generated within a dust collector. The five elements required for an explosion to occur are (1) fuel (dust), (2) oxygen (air), (3) dispersion (dust-laden air), (4) confinement (dust collector), and (5) ignition (typically a spark).

dynamic attack surface: The automated, on-the-fly changes of a building control system's characteristics to thwart actions of an adversary.

Dynamic Growth and Reconfiguration (DGR): A Dot Hill technology that allows the building control system administrator to quickly and easily add capacity or change RAID levels while the building control system is in use.

dynamic-link library: A Microsoft's implementation of shared library concept. Small software programs that are used when needed by another software program that is running on a Microsoft Windows computer that lets the large program communicate with a device such as a printer. Sometimes called a *DLL program* or *DLL file*.

Dynamic Routing Protocol: Used by a network device to learn communication routes such as when a router talks to adjacent routers to determine what networks each router is connected to.

dynamic subsystem: A subsystem that is not continually present during the execution phase of an information system. Service-oriented architectures and cloud computing architectures are examples of architectures that employ dynamic subsystems. SOURCE: SP 800-37

CHAPTER 6

E

Easter eggs: Benign message or joke generated by software hidden in a computer program or web page. Hidden functionality within an application program, which becomes activated when an undocumented, and often convoluted, set of commands and keystrokes are entered. Easter eggs are typically used to display the credits for the development team and are intended to be nonthreatening. SOURCE: SP 800-28

eavesdropping: Listening to a private conversation to acquire information that can provide access to a facility or BCS network.

echo reply: The response that a machine has received an echo request.

echo request: A message sent to a machine to determine if it is online and how long it takes to get to it.

egress filtering: Filtering outbound network traffic. SOURCE: SP 800-41

electric unit heater: Small heater used to prevent equipment freezing.

electrical power distribution: Four basic circuit arrangements are used for the distribution of electric power in buildings. They are: radial, primary selective, secondary selective, and secondary network circuit arrangements.

- **radial system:** The simplest and the lowest cost means of distributing the power when power is brought into a building at utilization voltage. There is only one utility source and all outgoing circuits feed loads. It is suitable for smaller installations where continuity of service is not critical.

- **primary selective system:** This circuit arrangement provides means of reducing both the extent and duration of an outage caused by a primary feeder fault through the use of duplicate primary feeder circuits and load interrupter switches that permit connection of each secondary substation transformer to either of the two primary circuits.

- **secondary selective system:** Under normal conditions, the secondary selective arrangement is operated as two separate radial systems. The secondary tie circuit breaker in each secondary substation is normally open.

- **secondary network system:** Many buildings with radial distribution systems are served at utilization voltage from utility secondary network systems. The network supply system assures a relatively high degree of service reliability.

SOURCE: Siemens

© Luis Ayala 2016
L. Ayala, *Cybersecurity Lexicon*, DOI 10.1007/978-1-4842-2068-9_6

electrical power usage report: An electrical power usage summary, operator selectable for substations, meters, or transducers, individual meters and transducers, any group of meters and transducers, and all meters for an operator selected time period. The report includes the voltage, current, power factor, electrical demand, and electrical power consumption, reactive power (Kvar) for each substation, facility, SCADA system, or equipment as selected by the operator. The report is automatically printed at the end of each summary period and includes:

- Total period consumption.

- Demand interval peak for the period, with time of occurrence.

- Energy consumption (kWh) over each demand interval.

- Time-of-use peak, semi-peak, off-peak, or baseline total kWh consumption.

- Reactive power during each demand interval.

- Power factor during each demand interval.

- Outside air (OA) temperature and relative humidity (RH) taken at the maximum and minimum of OA temperature of the report period with the time and dates of occurrence. At the installation's peak demand interval, the OA temperature and RH is also recorded.

SOURCE: Army Corps Guide Specification 13801

electromagnetic interference (EMI): What occurs when electromagnetic fields from one device interfere with the operation of another device.

Electronic Industries Association (EIA): A trade association that establishes electrical and electronics-oriented standards.

electronic vaulting: Transfer of data to an offsite facility using a telecommunication link to supplement full data backups.

elevator cyber-attack: A Trojan horse can be installed on a BCS network that when activated is able to put an elevator into diagnostics mode and send the elevator cab to another floor or cause it to stop between floors.

elevator surfing hack: A very dangerous activity whereby typically college students enter an elevator shaft and ride on top of elevators. Some even attempt to jump from the top of one moving elevator to another. Also known as *vator surfing* or *elevaroping*. Movement can also be provided by means of service controls located on top of the elevator car, which allow complete control over the movement of the elevator, but at a reduced speed.

elite: Cracker-speak meaning one of the cognoscenti, clueful, plugged-in. A hacker would be more likely to use the word *wizardly*.

emanations analysis: Obtaining data by monitoring and resolving electronic signals emitted by a building control system that contains the data, but is not designed to communicate the data.

embedded web server: Web server software built into a building control system device that is provided to configure the device from a web browser. Also used remotely by equipment vendors to update software or troubleshoot problems.

emergency: A sudden, unexpected event requiring immediate response due to possible threat to health and safety, or the environment.

emergency preparedness: The discipline that ensures an organization's readiness to respond in a coordinated, timely, and effective manner to a cyber-physical attack.

emergency procedures: A plan of action to prevent the loss of life and minimize injury and property damage after a cyber-attack.

encapsulation: Inclusion of one data structure within another structure so that the first data structure is hidden. Is an Object Oriented Programming concept that binds together the data and functions that manipulate the data, and that keeps both safe from outside interference and misuse. Data **encapsulation** led to the important OOP concept of data hiding. SOURCE: Wikipedia

encryption: Conversion of plaintext to ciphertext through the use of a cryptographic algorithm. The process of changing plaintext into ciphertext for the purpose of security or privacy. SOURCE: SP 800-21; CNSSI-4009 SOURCE: FIPS 185; CNSSI-4009

end-to-end encryption: Communications encryption in which data is encrypted when being passed through a network, but routing information remains visible. SOURCE: SP 800-12

end-to-end security: Safeguarding information in a building control system from point of origin to point of destination. SOURCE: CNSSI-4009

energy consumption control (ECC): The OFF switch.

energy control unit (ECU): A device that collects, processes and distributes lighting control information to control modules and wall controllers. Collects information from photo-sensors (light levels), occupancy sensors, and wall-mounted controllers.

Energy Management Controls System (EMCS): A building control system designed to enhance energy efficiency.

energy recovery unit (ERU): Mechanical equipment that includes a heat exchanger and a ventilation system for providing controlled ventilation into a building. The energy recovery ventilation process exchanges the energy contained in normally exhausted building or space air and uses it to precondition incoming outdoor air in a HVAC system.

enterprise: An organization with a defined mission/goal and a defined boundary, using information systems to execute that mission, and with responsibility for managing its own risks and performance. An enterprise may consist of all or some of the following business aspects: acquisition, program management, financial management (e.g., budgets), human resources, security, and information systems, information and mission management. SOURCE: CNSSI-4009

enterprise architecture (EA): The description of an enterprise's entire set of information systems: how they are configured, how they are integrated, how they interface to the external environment at the enterprise's boundary, how they are operated to support the enterprise mission, and how they contribute to the enterprise's overall security posture. SOURCE: CNSSI-4009

enterprise resource planning (ERP) system: A system that integrates enterprise-wide information including human resources, financials, manufacturing, and distribution as well as connects the organization to its customers and suppliers.

Enterprise Storage Network (ESN): An integrated suite of products and services designed to maximize heterogeneous connectivity and management of enterprise storage devices and servers; a dedicated, high-speed network connected to the enterprise's storage systems, enabling files and data to be transferred between storage devices and client mainframes and servers.

entrapment: Deliberate planting of apparent flaws in a BCS for the purpose of detecting attempted penetrations. SOURCE: CNSSI-4009

entropy: A measure of the amount of uncertainty that an attacker faces to determine the value of a secret. Entropy is usually stated in bits. SOURCE: SP 800-63

environment: The aggregate of external procedures, conditions, and objects that affect the development, operation, and maintenance of a building control system. SOURCE: FIPS 200; CNSSI-4009

environment of operation: The physical surroundings in which an information system processes, stores, and transmits information. SOURCE: SP 800-37; SP 800-53A The physical, technical, and organizational setting in which an information system operates, including but not limited to: missions/business functions; mission/business processes; threat space; vulnerabilities; enterprise and information security architectures; personnel; facilities; supply chain relationships; information technologies; organizational governance and culture; acquisition and procurement processes; organizational policies and procedures; organizational assumptions, constraints, risk tolerance, and priorities/trade-offs). SOURCE: SP 800-30

ephemeral port: Also called a *transient port* or a *temporary port*, usually on the client side. Typically set up when a client application wants to connect to a server and is destroyed when the client application terminates.

equipment schedule: Indicates the unique equipment identifier, manufacturer, model number, part number, and descriptive name for each control device, hardware, and component installed in a BCS.

error detection code: A code computed from data and comprised of redundant bits of information designed to detect, but not correct, unintentional changes in the data. SOURCE: FIPS 140-2; CNSSI-4009

escrow passwords: Passwords that are written down and stored in a secure location (like a safe) that are used by emergency personnel when privileged personnel are unavailable. SOURCE: FIPS 185

essential variable: A variable setpoint to be kept within preassigned limits.

Ethernet: A local area network standard for hardware, communication, and cabling. Also is the most widely installed local area network (LAN) technology. **Ethernet** is a link layer protocol in the TCP/IP stack, describing how networked devices can format data for transmission to other network devices on the same network segment, and how to put that data out on the network connection.

ethical hacker: A computer and networking expert who attempts to penetrate a building control system or network on behalf of its owners searching for security vulnerabilities that a malicious hacker could exploit.

event: An observable occurrence in a building control system or network. SOURCE: SP 800-61

event trigger: An event that causes another event to occur.

Evercookie: A JavaScript-based application that produces zombie cookies in a web browser that are intentionally difficult to delete. An Evercookie is not merely difficult to delete, it actively "resists" deletion by copying itself in different forms on the user's machine and resurrecting itself if it notices that some of the copies are missing or expired.

evil twin router: A fake wireless Internet hotspot set up to steal credit card information when victim connects.

examine: A type of assessment method that is characterized by the process of checking, inspecting, reviewing, observing, studying, or analyzing one or more assessment objects to facilitate understanding, achieve clarification, or obtain evidence, the results of which are used to support the determination of security control effectiveness over time. SOURCE: SP 800-53A

exculpatory evidence: Evidence that decreases the likelihood of fault or guilt.

exfiltration: Unauthorized transfer of information via a building control system.

explicit messaging: A proprietary vendor method of communication between devices where each message contains a message code that identifies the type of message and determines the action to be taken when received.

exploit: A technique to breach the security of a network or building control system in violation of security policy. An exploit cyber-attack is basically software designed to take advantage of a flaw in the building control system. The attacker plans to gain easy access to a building control system and gain control, allows privilege escalation, or creates a DOS attack. SOURCE: NICCS

exploit code: A program that allows attackers to automatically break into a building control system. SOURCE: SP 800-40

exploitable channel: Channel that allows the violation of the security policy governing a building control system and is usable or detectable by subjects external to the trusted computing base. SOURCE: CNSSI-4009

exploit kits: Easy to use software tools which cyber criminals buy that enables them to gain control of a computer. These are upgraded just like normal software the only difference is use of these kits is illegal.

exploitation analysis: In cybersecurity work, a person that analyzes collected information to identify vulnerabilities and the potential for exploitation. SOURCE: NICCS

exponential backoff algorithm: An algorithm used to adjust TCP timeout values so that BCS devices don't continue to timeout data.

exposure: A threat action whereby sensitive data is released to an unauthorized entity.

extended ACLs: A powerful form of Standard ACLs on Cisco routers. They make filtering decisions based on IP addresses (source or destination), ports (source or destination), protocols, and whether a session is established.

extended outage: A lengthy, unplanned interruption in building control system due to hardware or software problem or a communication failure.

Extensible Authentication Protocol (EAP): An authentication framework frequently used in wireless networks and point-to-point connections. It is defined in RFC 3748.

Extensible Markup Language (XML): A markup language that defines a set of rules for encoding documents in a format that is both human-readable and machine-readable. It is defined by the W3C's XML 1.0 Specification.

Exterior Gateway Protocol (EGP): The protocol that distributes routing information between two neighbor gateway routers that make up an autonomous building control system.

external interface file (XIF): A file which documents a device's external interface, specifically the number and types of LonMark objects, the number, types, directions, and connection attributes of network variables, and the number of message tags. SOURCE: Army Corps Spec 2015951

extra expense coverage: Insurance coverage for cyber-attack related expenses that may be incurred until operations are fully recovered after a cyber-attack.

extraction resistance: Capability of crypto-equipment or secure telecommunications equipment to resist efforts to extract keys. SOURCE: CNSSI-4009

extranet: A private network that uses web technology, permitting the sharing of portions of an enterprise's information or operations with suppliers, vendors, partners, customers, or other enterprises. SOURCE: CNSSI-4009

CHAPTER 7

F

facilities: Buildings designed and constructed for a specific function that contain the equipment, supplies, and voice and data communication lines, and require a building control system to operate under normal conditions.

facility (security) clearance (FCL): An FCL is an administrative determination that a company is eligible for access to classified information or award of a classified contract. Contract award may be made prior to the issuance of an FCL. The FCL requirement for a prime contractor includes those instances in which all classified access will be limited to subcontractors. Contractors are eligible for custody (possession) of classified material if they have an FCL and storage capability approved by the CSA. An administrative determination that a company is eligible for access to classified information up to a certain category. SOURCE: NISPOM 2-100

facility security officer (FSO): Provides operational oversight of a cleared facility's compliance with the requirements of the National Industrial Security Program (NISP).

fail-safe: Automatic protection of programs and/or processing systems when hardware or software failure is detected. SOURCE: CNSSI-4009

fail soft: Selective termination of affected nonessential processing when hardware or software failure is determined to be imminent. SOURCE: CNSSI-4009

failover: The capability to switch over automatically (typically without human intervention or warning) to a redundant or standby control system upon the failure or abnormal termination of the previously active building control system. SOURCE: SP 800-53; CNSSI-4009

failover protection: The transfer of operation from a failed component (e.g., controller, disk drive, pump) to a redundant component to ensure uninterrupted building equipment operations.

failure access: Type of incident in which unauthorized access to data results from hardware or software failure. SOURCE: CNSSI-4009

failure alarms: A failure alarm that combines inputs from multiple alarm points into a single, software-configured alarm. A low battery is a *minor* alarm. Low battery *and* an AC power failure *or* generator failure is a *major* alarm. A low battery *and* AC power failure *and* backup generator failure is a *critical* alarm.

failure control: Methodology used to detect imminent hardware or software failure and provide fail safe or fail soft recovery. SOURCE: CNSSI-4009

fake Flappy Bird attack: Malicious clones of mobile game apps such as *Flappy Bird* containing malware downloaded from app stores.

false acceptance: When a biometric system incorrectly identifies an individual or incorrectly verifies an impostor against a claimed identity. SOURCE: SP 800-76

false acceptance rate (FAR): The probability that a biometric system will incorrectly identify an individual or will fail to reject an impostor. The rate given normally assumes passive impostor attempts. SOURCE: SP 800-76

false flag: Covert operations designed to deceive in such a way that the operations appear as though they are being carried out by entities, groups, or nations other than those who actually planned and executed them. Also called a *black flag*.

false flame: Boiler error message that signifies a failed or leaky gas valve or flame detector malfunction.

false positive: An alert that incorrectly indicates that malicious activity is occurring. SOURCE: SP 800-61

false rejection: When a biometric system fails to identify an applicant or fails to verify the legitimate claimed identity of an applicant. SOURCE: SP 800-76

false rejection rate (FRR): The probability that a biometric system will fail to identify an applicant, or verify the legitimate claimed identity of an applicant. SOURCE: SP 800-76

fan coil unit (FCU): A simple device consisting of a heating or cooling coil and electric fan. Not connected to ductwork and only controls the temperature in the space where it is installed by a thermostat or a manual ON/OFF switch. More economical to install than ducted or central heating systems with air handling units.

fan cycling: A technique used to prevent cooling towers from freezing during cold weather. Fans are cycled a maximum of six on/off cycles per hour.

fan-powered terminal unit: A device that make use of the "free" heat that collects in the ceiling plenum after being emitted by lighting, people, and other equipment. Also called a *VAV box*. In a *parallel flow* terminal, the fan is outside the primary airstream and runs intermittently, (when the primary air is off). In a *series flow* terminal, the fan is in the primary airstream and runs constantly when the zone is occupied.

fan system effect: The negative impact on electric fan performance that results when it is installed in a duct. Fan performance is tested by laboratory measurements with the fan installed under ideal conditions. Fan performance is typically less than its rating under real world conditions. When electric fans are tested for rating, they are tested in an installation type that is common for the fan in question (i.e., ducted inlet, free outlet or ducted inlet, ducted outlet). Fan pressure changes are proportional to the square of a change in motor speed.

fast flux: A DNS technique used by botnets to hide phishing and malware delivery sites behind an ever-changing network of compromised hosts acting as proxies.

fault line attacks: Exploit gaps in coverage between interfaces of building control systems.

fault out of range: Detects whether the monitored value is outside the range of values considered to be normal for the object.

fault tolerance: The property that enables a system to continue operating properly in the event of the failure of (or one or more faults within) some of its components. If its operating quality decreases at all, the decrease is proportional to the severity of the failure, as compared to a naively designed system in which even a small failure can cause total breakdown. Fault tolerance is particularly sought after in high-availability or life-critical systems.

Federal Information Processing Standard (FIPS): A standard for adoption and use by federal departments and agencies that has been developed within the Information Technology Laboratory and published by the National Institute of Standards and Technology, a part of the US Department of Commerce. A FIPS covers some topic in information technology in order to achieve a common level of quality or some level of interoperability. SOURCE: FIPS 201

Federal Information System: An information system used or operated by an executive agency, by a contractor of an executive agency, or by another organization on behalf of an executive agency. SOURCE: SP 800-53; FIPS 200; FIPS 199; 40 U.S.C., Sec. 11331; CNSSI-4009

feeder: Single connection to the commercial power grid.

fiber channel (FC): A high-speed network technology (commonly running at 2-, 4-, 8- and 16-gigabit per second rates) primarily used to connect computer data storage.

fiber channel-arbitrated loop (FC-AL): A fiber channel topology in which devices are connected in a one-way loop fashion in a ring topology common within data storage systems.

Fiber Channel Community (FCC): An international non-profit organization whose members include manufacturers of servers, disk drives, RAID storage systems, switches, hubs, adapter cards, test equipment, cables and connectors, and software solutions.

Fiber Distributed Data Interface (FDDI): A 100 Mbit/s ANSI standard LAN architecture, defined in X3T9.5. The underlying medium is optical fiber (though it can be copper cable, in which case it may be called CDDI) and the topology is a dual-attached, counter-rotating token ring.

Fieldbus: A digital, serial, multi-drop, two-way data bus or communication path or link between low-level industrial field equipment such as sensors, transducers, actuators, local controllers, and even control room devices. Use of fieldbus technologies eliminates the need of point-to-point wiring between the controller and each device. A protocol is used to define messages over the fieldbus network with each message identifying a particular sensor on the network. SOURCE: SP 800-82

field control system (FCS): A building control system or utility control system.

field device hack: Equipment that is connected to the field side on a ICS. Types of field devices include RTUs, PLCs, HMIs, actuators, sensors, and associated communications. All can be hacked.

field site: A subsystem identified by physical, geographical, or logical segmentation within the ICS. A field site may contain RTUs, PLCs, actuators, sensors, HMIs, and associated communications. SOURCE: SP 800-82

Field Transfer Protocol (FTP): Internet standard for transferring files over the Internet. Used to download and upload web pages, graphics and other files between local media and a remote server. FTP is a program that is shipped with most of the Windows, Unix, and Linus operating systems.

file backup: The practice of copying a file stored on disk to another disk, thumb drive, tape or CD in case the active file is damaged.

file infector virus: A virus that attaches itself to a program file, such as a word processor, spreadsheet application, or game. SOURCE: SP 800-61

file integrity monitoring (FIM): Host-based intrusion detection software installed on an asset that analyzes building control system behavior and configuration status to track user access and activity as well as identify potential security exposures such as

- building control system compromise

- modification of critical configuration files (e.g., registry settings, password, etc.)

- common rootkits

- rogue processes

file name anomaly:

- A mismatch between the internal file header and its external extension; or

- A file name inconsistent with the content of the file (e.g., renaming a graphics file with a non-graphical extension.)

SOURCE: SP 800-72

file protection: Aggregate of processes and procedures designed to inhibit unauthorized access, contamination, elimination, modification, or destruction of a file or any of its contents. SOURCE: CNSSI-4009

file recovery: Restoration of computer files using backup copies.

file-renaming tricks: Hackers sometimes use file name tricks to get victims to execute malicious code such as naming the file something that would encourage unsuspecting victims to click on it. Microsoft Windows readily hides common file extensions, so a file named NudePics.Gif.exe is displayed as NudePics.Gif.

file server: Central repository of shared files and applications in a building controls system.

finger: A protocol to look up user information on a host using an e-mail address as input and return information about the user who owns that e-mail address. Some building control systems provide the user's full name, address, and telephone number.

FIREFLY: Key management protocol based on public key cryptography. SOURCE: CNSSI-4009

fire sale attack: As seen in the 2007 movie *Live Free or Die Hard*. A Fire Sale is a plan to create chaos and an assault against a government, transportation, and economy by computer hackers. Much like the more common term meaning "everything must go." Any and all computer-based systems are the objective and would destroy the modern-day life of a nation. The three stages of a fire sale in the movie are:

> **Stage 1:** Shut down transportation systems: traffic lights, railroads, subways, and airports.

> **Stage 2:** Disable financial systems: Wall Street, banks and financial records.

> **Stage 3:** Turn off public utility systems: electricity, gas lines, telecom and satellite systems.

firewall: A hardware/software capability that limits access between networks and/or systems in accordance with a specific security policy. SOURCE: CNSSI-4009

firewall control proxy: The component that controls a firewall's handling of a call. The firewall control proxy can instruct the firewall to open specific ports that are needed by a call, and direct the firewall to close these ports at call termination. SOURCE: SP 800-58

firmware: The programs and data components of a cryptographic module that are stored in hardware within the cryptographic boundary and cannot be dynamically written or modified during execution. SOURCE: FIPS 140-2

FitBit hack: FitBit is shown as hacked, creating a possible cyber-attack vector that could allow hackers to infect any PC connected to it. A virus is uploaded to the FitBit that uploads to your PC via USB.

Flame virus: This computer virus can record audio, screenshots, keyboard activity, and network traffic. It can record Skype conversations and can turn infected computers into Bluetooth beacons that attempt to download contact information from nearby Bluetooth-enabled devices. Also known as **Flamer**, **Da Flame**, **sKyWIper**, and **Skywiper**. Flame supports a "kill" command that wipes all traces of the malware from the computer. Due to the size and complexity of the program (20 MB), it is described as "twenty times" more complicated than Stuxnet.

flame check error: BCS error message that ionization (flame/rod) wire lost signal for more than 15 seconds.

flaws and features: A flaw is *unintended functionality*. This may either be a result of poor design or through mistakes made during implementation. Flaws may go undetected for a significant period of time. The majority of common attacks today exploit these types of vulnerabilities. A feature is *intended functionality* that can be misused by an attacker to breach a building control system. Features may improve the user's experience, help diagnose problems or improve management, but they can also be exploited by an attacker. Macros are examples of features that can be misused by a hacker.

flaw: Error of commission, omission, or oversight in a building control system that may allow protection mechanisms to be bypassed. SOURCE: CNSSI-4009

Flaw Hypothesis Methodology: System analysis and penetration technique in which the specification and documentation for a building control system are analyzed to produce a list of hypothetical flaws. This list is prioritized on the basis of the estimated probability that a flaw exists, on the ease of exploiting it, and on the extent of control or compromise it would provide. The prioritized list is used to perform penetration testing of a building control system. SOURCE: CNSSI-4009

flick warning: Prior to a scheduled electric outage or expiry of a temporary override, the building control system provides two short light level drops as a warning to the affected building occupants.

flooding attack: Cyber-attack that attempts to cause a failure in the security of a building control system or industrial control device by providing more input than the device can process properly. SOURCE: CNSSI-4009

flow meter: Inline electromagnetic flow meters are suitable for measurement of electrically conductive liquids. Inherently bi-directional, a flow meter is equipped with a standard transmitter that provides output for flow rate and programmable outputs.

flow switch: For starting or stopping electrically operated equipment such as signal lights, alarms, motors, automatic burners, and metering devices.

focused testing: A test methodology that assumes some knowledge of the internal structure and implementation detail of the assessment object. Also known as *gray box testing*. SOURCE: SP 800-53A

footprint: The amount of floor space that a piece of equipment (e.g., a rackmount enclosure) occupies.

forensic copy: An accurate bit-for-bit reproduction of the information contained on an electronic device or associated media, whose validity and integrity has been verified using an accepted algorithm. SOURCE: SP 800-72; CNSSI-4009

forensic specialist: A professional who locates, identifies, collects, analyzes, and examines data while preserving the integrity and maintaining a strict chain of custody of information discovered. SOURCE: SP 800-72

forensics: The practice of gathering, retaining, and analyzing computer-related data for investigative purposes in a manner that maintains the integrity of the data. SOURCE: CNSSI-4009

forensically clean: Digital media that is completely wiped of all data, including nonessential and residual data, scanned for malware, and verified before use. SOURCE: SP 800-86

fork bomb attack: A cyber-attack that works by using the fork() call to create a new process which is a copy of the original. By doing this repeatedly, all available processes on the machine can be taken up.

form factor: The physical size and shape of a device; often used to describe the size of hard drive arrays in a rack mount enclosure.

forward recovery: Recovering a building control system to the point of failure by applying active data to current backup files of the database.

Foreign Security Service (FSS): Intelligence agency of a foreign nation.

fragment overlap attack: The IP fragment overlapped exploit occurs when two fragments contained within the same IP datagram have offsets that indicate that they overlap each other in positioning within the datagram. This could mean that either fragment A is being completely overwritten by fragment B, or that fragment A is partially being overwritten by fragment B. Some operating systems do not properly handle fragments that overlap in this manner and may throw exceptions or behave in other undesirable ways upon receipt of overlapping fragments. Overlapping fragments may be used in an attempt to bypass Intrusion Detection Systems. In this exploit, part of an attack is sent in fragments along with additional random data; future fragments may overwrite the random data with the remainder of the attack. If the completed datagram is not properly reassembled at the IDS, the cyber-attack will go undetected.

freeze stat: A temperature-sensing device for HVAC that monitors a heat exchanger to prevent its coils (air coil or liquid coil) from freezing.

frequency converter: An electronic device that converts alternating current of one frequency to another frequency and may also change the voltage. Frequency converters are typically used to control the speed of motors, primarily pumps and fans on industrial processing lines, where the control accuracy requirements can be very high. This is particularly useful in the nuclear power and weapons industry, where these devices regulate the operation of refinement centrifuges. The STUXNET worm targeted specific Siemens brand frequency converters, causing erratic operation.

frequency hopping: Repeated switching of frequencies during radio transmission according to a specified algorithm, to minimize unauthorized interception or jamming of telecommunications. SOURCE: CNSSI-4009

friction stir welding (FSW): A solid-state joining process (the metal is not melted) that joins two facing surfaces.

fuel cell: A device that converts the chemical energy from a fuel into electricity through a chemical reaction of positively charged hydrogen ions with oxygen or another oxidizing agent. Keep in mind that a fuel cell large enough to power an entire building (or a critical components) is generally not a substitute for a backup generator because a fuel cell typically needs a reference voltage. So, when the commercial power is unavailable the reference voltage is gone and the fuel cell will shut down. It's also important to remember that a fuel cell that runs on natural gas may not operate properly if the utility company adds propane to the system such as when natural gas supplies are low.

full-disk encryption (FDE): The process of encrypting all the data on the hard disk drive used to boot a computer, including the computer's operating system, and permitting access to the data only after successful authentication with the full disk encryption product. SOURCE: SP 800-111

full duplex: Refers to the transmission of data in two directions simultaneously (sender and receiver can send at the same time).

full maintenance: Complete diagnostic repair, modification, and overhaul of COMSEC equipment, including repair of defective assemblies by piece part replacement. See Limited Maintenance. SOURCE: CNSSI-4009

full recovery test: An exercise in which all recovery procedures are tested.

function pointer attack: A buffer overflow by overwriting a function pointer or exception handler, which is subsequently executed.

functional profile: A standard description of one or more LonMark Objects used to classify and certify devices.

functional testing: Segment of security testing in which advertised security mechanisms of an information system are tested under operational conditions. SOURCE: CNSSI-4009

fuzzing attack: A cyber-attack when indiscriminate data is transmitted to a server in an attempt to override controls.

CHAPTER 8

G

gas utility flow meter: A specialized flow meter used to measure the volume of fuel gases such as natural gas and propane. Gases are more difficult to measure than liquids, as measured volumes are highly affected by temperature and pressure. Gas meters measure a defined volume, regardless of the pressurized quantity or quality of the gas flowing through the meter. Temperature, pressure, and heating value compensation must be made to measure actual amount and value of gas moving through a meter. Several different designs of gas meters are in common use, depending on the volumetric flow rate of gas to be measured, the range of flows anticipated, the type of gas being measured, and other factors.

gateway: Interface providing compatibility between networks by converting transmission speeds, protocols, codes, or security measures. SOURCE: CNSSI-4009

Gauss malware: Malware able to steal bank account credentials and gather as much information about the infected machines as possible. It is believed the purpose is likely espionage, not theft.

general purpose programmable controller (GPPC): Unlike an ASC or AGC, a GPPC is not furnished with a fixed application program and does not have a fixed ProgramID or XIF file. A GPPC can be reprogrammed. SOURCE: UFGS 25 10 10

general support system: An interconnected set of information resources under the same direct management control that shares common functionality. It normally includes hardware, software, information, data, applications, communications, and people. An interconnected set of information resources under the same direct management control which shares common functionality. A system normally includes hardware, software, information, data, applications, communications, and people. A system can be, for example, a local area network (LAN) including smart terminals that supports a branch office, an agency-wide backbone, a communications network, a departmental data processing center including its operating system and utilities, a tactical radio network, or a shared information processing service organization (IPSO). SOURCE: OMB Circular A-130, App. III; CNSSI-4009

generator: An independent source of on-site power usually fueled by diesel oil, natural gas, or both.

geotagging: The process of adding geographical identification metadata to various media such as a geotagged photograph or video, web sites, SMS messages, QR Codes or RSS feeds and is a form of geospatial metadata. This data usually consists of latitude and longitude coordinates, though they can also include altitude, bearing, distance, accuracy data, and place names.

gethostbyaddr: DNS query when the address of a machine is known and the name is needed.

gethostbyname: DNS quest when the name of a machine is known and the address is needed.

ghost apps: A software program typically on a cell phone or tablet that is designed to help users hide their activity and camouflage sensitive information. A Ghost App is disguised to look like a normal app such as a calculator, but once you type in a secret code, it takes you to a hidden page where you can store photos, video, and all kinds of personal information. Also called a Vault App or Photo Vault. Some vault apps

© Luis Ayala 2016
L. Ayala, *Cybersecurity Lexicon*, DOI 10.1007/978-1-4842-2068-9_8

specialize in hiding photos, but others allow users to filter text messages and hide phone calls. Some Vault Apps use a decoy password so when someone sees the photo vault on your phone, you tell them the decoy password and it opens up to entire set of fake secret photos so people lose their curiosity.

ghostware: "Stealth" programs—usually for monitoring, like Trojans, keyloggers, and so forth—that reside in a system and are not readily detectible by the user. They transmit information to the person that installed the programs without the PC user being able to tell that it's there. Software designed to rid a system of adware, viruses, and the like, may not be able to tell if ghostware is on a PC. SOURCE: Microsoft Research, Redmond.

gigabyte: Approximately one billion bytes, or 1,024 megabytes.

glob: In computer programming, in particular in a Unix-like environment, glob patterns specify sets of file names with wildcard characters. For example, the Unix command mv *.txt textfiles/ moves (mv) all files with names ending in .txt from the current directory to the directory textfiles. Here, * is a wildcard standing for "any string of characters" and *.txt is a glob pattern. The other common wildcard is the question mark (?), which stands for one character. The operation of matching of wildcard patterns to multiple file or path names is referred to as *globbing*.

global information grid (GIG): The globally interconnected, end-to-end set of information capabilities for collecting, processing, storing, disseminating, and managing information on demand to warfighters, policy makers, and support personnel. The GIG includes owned and leased communications and computing systems and services, software (including applications), data, security services, other associated services, and national security systems. Non-GIG IT includes stand-alone, self-contained, or embedded IT that is not, and will not be, connected to the enterprise network. SOURCE: CNSSI-4009

global information infrastructure (GII): Worldwide interconnections of the information systems of all countries, international and multinational organizations, and international commercial communications. SOURCE: CNSSI-4009

Gnutella: Internet file sharing utility.

GoogleDiggity: Traditional Google hacking tool.

Google Dork: An inept or foolish person as revealed by Google.

GoogleDorks: Google search queries that uncover vulnerable building control systems and/or sensitive information disclosures listed in the original Google Hacking Database (GHDB).

Google hacking: A computer hacking technique that uses Google Search and other Google applications to find security holes in the configuration and computer code that web sites use. Google hacking involves using advanced operators in the Google search engine to locate specific strings of text within search results. Some of the more popular examples are finding specific versions of vulnerable web applications. A search query could locate all web pages that have particular text contained within them. Some searches can even retrieve the username and password list from Microsoft FrontPage servers.

Google Hacking Database (GHDB): A dictionary listing of automated attack tools to find vulnerable building control systems and sensitive information disclosures in public building control systems that have been indexed by search engines. SOURCE: SP 800-30

Google Hacking Diggity Project: A research and development initiative dedicated to investigating the latest techniques that leverage search engines (such as Google, Bing, and Shodan) to quickly identify vulnerable building control systems and sensitive data on public networks. Free attack and defense tools related to search engine hacking are available for download.

Govware: In German-speaking countries, spyware used or made by the government is sometimes called *govware*. Govware is typically a Trojan horse used to intercept communications from the target computer. Some countries like Switzerland and Germany have a legal framework governing the use of such software. In the US, the term *policeware* has been used for similar purposes.

graduated security: A security system that provides several levels (e.g., low, moderate, high) of protection based on threats, risks, available technology, support services, time, human concerns, and economics. SOURCE: FIPS 201

gray hole attack: A type of packet drop attack in which a router that is supposed to relay packets instead discards them for a particular network destination, at a certain time of the day, a packet every n packets or every t seconds, or a randomly selected portion of the packets. This usually occurs from a router becoming compromised from a number of different causes. Because packets are routinely dropped from a lossy network, the packet drop attack is very hard to detect and prevent.

grid independent: A facility that produces sufficient energy on site and does not rely on commercial electricity.

grid-interactive inverters: Electrical devices that convert direct current (DC) to alternating current (AC) that have the additional requirement that they produce AC power that matches the existing power presented on the grid. In particular, a grid-interactive inverter must match the voltage, frequency, and phase of the power line it connects to.

graphical system display: The graphical display consists of building control system (air handler units, VAV boxes, chillers, cooling towers, boilers, etc.) graphic displays. Data associated with an active display is updated every 5 seconds. Each building or building subarea display shows the building foot print and basic floor plan, and distinguishes between the individual zones and the equipment serving each zone and space. The building display shows all space sensor and status readings, as applicable, for the individual zones such as space temperature, humidity, occupancy status, and so forth. The building display also shows the locations of individual pieces of monitored and controlled equipment. Also referred to as a *graphical user interface.*

grep: Command-line tool for searching plain-text data sets for lines matching a regular expression. Originally developed for the Unix operating system (global regular expression print), but is available today for all Unix-like systems. Grep can be used to find the URLs that apps are programmed to contact. A simple example of a common usage of grep is the following, which searches the file fruitlist.txt for lines containing the text string apple:

```
$ grep apple fruitlist.txt
```

ground plane: An electrically conductive surface, usually connected to electrical ground.

groupware: Application software designed to help people involved in a common task to achieve their goals. One of the earliest definitions of collaborative software is "intentional group processes plus software to support them." SOURCE: Wikipedia

GSM base station: A cellular network. Cell phones connect to it by searching for cells in the immediate vicinity. The longest distance GSM supports is 22 miles.

guessing entropy: A measure of the difficulty that a hacker has to guess the average password used in a building control system. Entropy is typically stated in bits. When a password has n-bits of guessing entropy then a hacker has as much difficulty guessing the average password as in guessing an n-bit random quantity. The hacker is assumed to know the actual password frequency distribution. SOURCE: SP 800-63

CHAPTER 9

H

hack attack map: A web site that shows where cyber-attacks are coming from and who is being attacked, in real time. A good example is map.ipviking.com. Also called a *cyber-threat map*.

hackback: Vigilante attempt to retaliate after a hacker attacks your system. Definitely still illegal, even if you know for sure who is responsible.

HackRF: A software defined radio. Similar to BladeRF and USDR. *See Universal Software Radio Peripheral (USRP).*

hacker: Unauthorized user who attempts to or gains access to an information system. SOURCE: CNSSI-4009 Also written as *hax0r*. The hacker can be any one of the following:

- **Loner:** Single individual, can be a hacktivist, extremist or criminal. Domestic or foreign. Motivation could be for shits and grins, nuisance, revenge or merely a public display of capability (bragging rights).

- **Insider Threat:** Disgruntled employee or Contractor personnel, male or female.

- **Foreign Nation-State**

- **Terrorist Organization**

- **Criminal Organization**

hackerspace: A community-operated workspace where people with common interests, often in computers, machining, technology, science, digital art, or electronic art, can meet, socialize, and collaborate. Hackerspaces have also been compared to other community-operated spaces with similar aims and mechanisms such as Fab Lab, Men's Sheds, and commercial for-profit companies such as TechShop (also referred to as a *hacklab, makerspace*, or *hackspace*).

hacking: Utilizing features of something to do an unexpected and unintended thing.

hacktivist: The subversive use of computers and computer networks to promote a political agenda.

handshaking procedures: Dialog between two information systems for synchronizing, identifying, and authenticating themselves to one another. SOURCE: CNSSI-4009

halon: A toxic gas used to extinguish fires effective only in closed areas. It is being phased out.

Hall-effect water meter: Solid-state flow detection device.

hand mode: Equipment is operated manually through a keypad located at the device, independent of the BCS. Also referred to as *local mode*.

© Luis Ayala 2016
L. Ayala, *Cybersecurity Lexicon*, DOI 10.1007/978-1-4842-2068-9_9

Hand-Off-Auto switch (H-O-A): A device that has switches that maintain their position. Start and Stop buttons have momentary actions. The Off position will prevent any operation. Used in a situation that has a single point of manual control to allow the motor (or other device) to (A) operate from an automated building control system, (O) not operate, or (H) operate with no safeguards or automated control. The Hand position is used to bump the motor or to operate for shorts times while being observed by operating personnel.

hardening: Configuring a host's operating systems and applications to reduce the host's security weaknesses. SOURCE: SP 800-123

hard zero: When a hard zero is performed during equipment calibration then any adjustments will be maintained when the device is turned off.

hashcash: A proof-of-work algorithm used to limit e-mail spam and denial-of-service attacks that requires a selectable amount of work to compute, but the proof can be verified efficiently. For e-mail uses, a textual encoding of a hashcash stamp is added to the header of an e-mail to prove the sender has expended a modest amount of CPU time calculating the stamp prior to sending the e-mail. As the sender has taken a certain amount of time to generate the stamp and send the e-mail, it is unlikely that they are a spammer. The receiver verifies the stamp is valid.

hash function: A function that maps a bit string of arbitrary length to a fixed length bit string. Approved hash functions satisfy the following properties:

- One-Way. It is computationally infeasible to find any input that maps to any prespecified output.

- Collision Resistant. It is computationally infeasible to find any two distinct inputs that map to the same output.

SOURCE: SP 800-63; FIPS 201

hash total: Value computed on data to detect error or manipulation. SOURCE: CNSSI-4009

hash value: The result of applying a cryptographic hash function to data (e.g., a message). SOURCE: SP 800-106

hashing: A process of applying a mathematical algorithm against a set of data to produce a numeric value (a "hash value") that represents the data. SOURCE: SP 800-72; CNSSI-4009

headless worm: Although the Conficker computer worm that used to command a large botnet was neutralized years ago prevented the people who released the worm from using it, the malicious code can still be found on infected personal computers.

heartbeat signals: Communications traffic that signals the health of the building control system. It is also referred to as a *watchdog timer*, *keep-alive*, or *health status*.

heat load: Total heat per unit time needed in order to maintain a specified temperature in a building.

heating, ventilation, and air conditioning (HVAC): The mechanical equipment that provides and maintains a controlled environment with conditions conducive to continuous and uninterrupted building operations.

Hi-Link network devices: Hi-Link network devices are being designed to understand all the languages that connected devices use and be able to communicate with them in their own language. This approach is an attempt to unify the languages in which intelligent electronic devices communicate, but at the network level and not the device level. The Hi-Link protocol was developed by Huawei to unify the *Internet of Things*.

hidden apps: A software program typically on a cell phone or tablet that is designed to help users hide their activity and camouflage sensitive information. A hidden app is disguised to look like a normal app such as a calculator, but once you type in a secret code, it takes you to a hidden page where you can store photos, video, and personal information. Some hidden apps specialize in hiding photos, but others allow users to filter text messages and hide phone calls. Some hidden apps use a decoy password so when someone discovers the hidden app on your phone, you tell them the decoy password and it opens up to an entire set of "fake" secret data so people lose their curiosity. Also called a *ghost app, vault app*, or *photo vault*.

hierarchical storage management (HSM): a data storage technique, which automatically moves data between high-cost and low-cost storage media. HSM systems exist because high-speed storage devices, such as hard disk drive arrays, are more expensive (per byte stored) than slower devices, such as optical discs and magnetic tape drives.

high-assurance guard (HAG): An enclave boundary protection device that controls access between a local area network that an enterprise system has a requirement to protect, and an external network that is outside the control of the enterprise system, with a high degree of assurance. A guard that has two basic functional capabilities: a *message guard* and a *directory guard*. The message guard provides filter service for message traffic traversing the guard between adjacent security domains. The directory guard provides filter service for directory access and updates traversing the guard between adjacent security domains. SOURCE: SP 800-32; CNSSI-4009

high availability: A failover feature to ensure availability during device or component interruptions. SOURCE: SP 800-113

high impact: The loss of confidentiality, integrity, or availability that could be expected to have a severe or catastrophic adverse effect on organizational operations, organizational assets, individuals, other organizations, or the national security interests of the United States. (1) causes a severe degradation in mission capability to an extent and duration that the organization is able to perform its primary functions, but the effectiveness of the functions is significantly reduced; (2) results in major damage to organizational assets; (3) results in major financial loss; or (4) results in severe or catastrophic harm to individuals involving loss of life or serious life-threatening injuries. SOURCE: FIPS 199; CNSSI-4009

high-level alarm: A float level sensor that provides an audible alarm for a high fluid level, low fluid level, or serves as an overfill or interstitial alarm. The switch typically must be manually reset upon correcting the alarm condition.

high-temperature alarm: High- and low-temperature alarm alerts operator with audio and visual alarms when temperature is outside predetermined operating range.

high-priority tasks: Activities vital to the operation of the organization.

hijack attack: Active wiretapping in which a hacker seizes control of a previously established communication connection.

historian: A building control system computer that stores values for various processes or states of interest to the building control system. They are often the point of connection between the corporate network and the building control system network.

hit inflation attack: A hit inflation attack is a kind of fraudulent method used by some Internet advertisement publishers to earn unjustified revenue on the click traffic they drive to the advertisers' web sites. It is more sophisticated and harder to detect than a simple inflation attack.

HoneyMonkey: Automated system simulating a user browsing web sites. Used to detect which web sites exploit vulnerabilities in the browser. Also known as *HoneyClient*.

Honeynet Project: An international security research organization, "dedicated to investigating the latest attacks, developing open source security tools to improve Internet security and learning how malicious hackers behave."

Honeypot: A system (e.g., a web server) or system resource (e.g., a file on a server) that is designed to be attractive to potential crackers and intruders and has no authorized users other than its administrators. SOURCE: CNSSI-4009

home page: The main page on a web site that serves as the primary point of entry to related pages within the site and may have links to other sites as well.

home security systems: Historically, home security systems were hardwired to a service provider operations center, however today they are connected wirelessly. This allows a homeowner to monitor his home at all times, from anywhere, using a smartphone or tablet. It also makes it much easier to be hacked.

host-attached storage: A storage system that is connected directly to the network server; also referred to as *server-attached storage*.

host-based intrusion detection system (HIDS): Detects malicious activity on a host from characteristics such as change of files (file system integrity checker) or operating system profiles.

host bus adapter (HBA): A hardware daughterboard that resides on the computer bus and provides an interface connection between a SCSI device (such as a hard drive) and the host computer.

hot site: A fully operational offsite data processing facility equipped with hardware and software, to be used in the event of an information system disruption. Backup site that includes phone systems with the phone lines already connected. Networks will also be in place, with any necessary routers and switches plugged in and turned on. Desks will have desktop PCs installed and waiting, and server areas will be replete with the necessary hardware to support business-critical functions. Within a few hours, a hot site can become a fully functioning element of an organization. SOURCE: SP 800-34 & CNSSI-4009

hot spare: A backup component (e.g., hard drive or controller) that is online and available in the event the primary component goes down.

Hotspot ID: Tracks Wi-Fi connections and warns users when they connect to an unsafe access point. A trademark of Red Dog Communications.

hot swappable: The ability to replace a component (e.g., defective I/O module, hard drive, controller, fan or power supply) while the building control system is on line, without having to power down; also referred to as *hot-plug removable*.

hot wash: A debrief conducted immediately after a cyber-exercise or test with the staff and participants. SOURCE: SP 800-84

hot-water heater hack: An electric heating element installed in a water storage tank (or it can be tankless). A tankless water heater only heats the water it is drawn. Other fuel choices include hybrid electric/heat-pump models, solar water heaters, and condensing gas water heaters. Hacking a water heater would allow an attacker to superheat the water exposing building occupants to dangerously hot water. Unfortunately, pressure relief valves have been known to fail and if the water heater becomes over-pressurized, it can explode with terrific force.

hot-water loop: Used to distribute hot water throughout a building. In a normal plumbing system there is normally one main hot water line. This line feeds branches that further split to feed the various hot water taps. In a hot water plumbing loop (also known as a *closed loop*) the line from the hot water system continues from one tap, to the next and back to the source. Typically, hot water circulates continuously in the loop so a person does not have to waste clean water waiting for hot water to reach the tap, thereby reducing water waste.

hub: A device that splits one network cable into a set of separate cables, each connecting to a different computer; used in a building control system to create a small-scale network.

human-machine interface (HMI): The computer hardware and software that enables a single operator to monitor and control equipment remotely.

human threats: Possible source of cyber-attack resulting from human actions (i.e., disgruntled employee, terrorism, etc.).

humidity/temperature sensor: Humidity transducers designed for use with HVAC direct digital controllers. Ceramic technology humidity sensors are typically not affected by condensation.

HVAC control system drawings: Facility drawings (and CAD files) that show the mechanical equipment schedules, equipment locations, equipment connections and Sequence of Operations that make up the building's HVAC system. The HVAC drawing legend indicates all symbols, abbreviations, and acronyms.

HVAC testing, adjusting and balancing (TAB): These are the three major steps in commercial building construction used to achieve proper operation of HVAC systems. Usually performed as part of the commissioning process. A specialist performs air and hydronic measurements on the HVAC systems and adjusts the flows as required to achieve optimum performance of the HVAC equipment.

hybrid cyber-attack: A cyber-attack that builds on the dictionary attack method by adding numerals and symbols to dictionary words.

hybrid security control: A security control that is implemented in an information system in part as a common control and in part as a system-specific control. SOURCE: SP 800-37; SP 800-53; SP 800-53A; CNSSI-4009

hydraulic motion: Hydraulic actuators lift and hold heavy loads without brakes, move heavy objects at slow speeds, apply torque without gearing. For positioning large loads, electro hydraulics can be a better choice than electromechanical motion systems. Hydraulic actuators need less space and produce less heat than electric motors.

hypervisor: A hypervisor or virtual machine monitor (VMM) is a part of computer software, firmware, or hardware that creates and runs virtual machines. A computer on which a hypervisor is running one or more virtual machines is defined as a host machine. Each virtual machine is called a *guest machine*. The hypervisor presents the guest operating systems with a virtual operating platform and manages the execution of the guest operating systems. Multiple instances of a variety of operating systems may share the virtualized hardware resources. Can be used to defend against certain types of cyber-attacks and is a potential vulnerability to other types of cyber-attacks.

CHAPTER 10

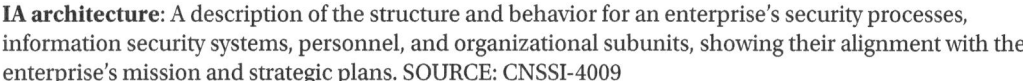

I

IA architecture: A description of the structure and behavior for an enterprise's security processes, information security systems, personnel, and organizational subunits, showing their alignment with the enterprise's mission and strategic plans. SOURCE: CNSSI-4009

IA infrastructure: The underlying security framework that lies beyond an enterprise's defined boundary, but supports its IA and IA-enabled products, its security posture and its risk management plan. SOURCE: CNSSI-4009

IA product: Product whose primary purpose is to provide security services (e.g., confidentiality, authentication, integrity, access control, non-repudiation of data); correct known vulnerabilities; and/ or provide layered defense against various categories of non-authorized or malicious penetrations of information systems or networks. SOURCE: CNSSI-4009

IA-enabled information technology product: Product or technology whose primary role is not security, but which provides security services as an associated feature of its intended operating capabilities. Examples include such products as security-enabled web browsers, screening routers, trusted operating systems, and security-enabled messaging systems. SOURCE: CNSSI-4009

IA-enabled product: Product whose primary role is not security, but provides security services as an associated feature of its intended operating capabilities. Examples include such products as security-enabled web browsers, screening routers, trusted operating systems, and security enabling messaging systems. SOURCE: CNSSI-4009

ICS manager: An individual who typically oversees industrial operations, focusing on product quality, environmental protection, and industrial safety. Creates and maintains automated building control systems that regulate temperature, lighting, humidity, water, and electricity as well as automated industrial control systems. Calibrates machines, troubleshoots equipment, and repairs or replaces instruments.

identification: The process that enables, generally by the use of unique machine-readable names, recognition of users or resources, as indistinguishable, to those previously described to the automated building control system. The process of verifying the identity of a user, process, or device, usually as a prerequisite for granting access to resources in an IT system. SOURCE: SP 800-47

identity: A set of attributes that uniquely describe a person within a given context. The set of physical and behavioral characteristics by which an individual is uniquely recognizable. The set of attribute values (i.e., characteristics) by which an entity is recognizable and that, within the scope of an identity manager's responsibility, is sufficient to distinguish that entity from any other entity. SOURCE: SP 800-63; FIPS 201; CNSSI-4009

identity-based access control: Access control based on the identity of the user (typically relayed as a characteristic of the process acting on behalf of that user) where access authorizations to specific objects are assigned based on user identity. SOURCE: SP 800-53; CNSSI-4009

identity-based security policy: A security policy based on the identities and/or attributes of the object (system resource) being accessed and of the subject (user, group of users, process, or device) requesting access. SOURCE: SP 800-33

identity registration: The process of making a person's identity known to the Personal Identity Verification (PIV) system, associating a unique identifier with that identity, and collecting and recording the person's relevant attributes into the system. SOURCE: FIPS 201; CNSSI-4009

identity theft: The deliberate use of someone else's identity, usually as a method to gain a financial advantage or obtain credit and other benefits in the other person's name. When this is done online on the Internet, it is called *online identity theft*. One of the most common types of cybercrime. The most common source to steal identity information of others, are data breaches affecting government web sites and private web sites.

identity token: Smart card, metal key, or other physical object used to authenticate identity for the use of a service. Such as an application or a product.

image: An exact bit-stream copy of all electronic data on a device, performed in a manner that ensures that the information is not altered. SOURCE: SP 800-72

imitative communications deception: Introduction of deceptive messages or signals into an adversary's telecommunications signals. SOURCE: CNSSI-4009

impact level: The magnitude of harm that can be expected to result from the consequences of unauthorized disclosure of information, unauthorized modification of information, unauthorized destruction of information, or loss of information or information system availability. SOURCE: CNSSI-4009 High, Moderate, or Low security categories of an information system established in FIPS 199 which classify the intensity of a potential impact that may occur if the information system is jeopardized. SOURCE: SP 800-34

impact value: The assessed potential impact resulting from a compromise of the confidentiality, integrity, or availability of an information type, expressed as a value of low, moderate, or high. SOURCE: SP 800-30

implant: Electronic device or electronic equipment modification designed to gain unauthorized interception of information-bearing emanations. SOURCE: CNSSI-4009

in real life (IRL): Internet users may speak of having "met" someone that they have contacted via online chat or in an online gaming context. To say that they met someone "in real life" is to say that they literally encountered them in a common physical location (AFK: Away From Keyboard).

incident: An occurrence that actually or potentially jeopardizes the confidentiality, integrity, or availability of a building control system or the information the building control system processes, stores, or transmits or that constitutes a violation or imminent threat of violation of security policies, security procedures, or acceptable use policies. Incidents may be intentional or unintentional. SOURCE: FIPS 200; SP 800-53

incident detection: Ways in which a cyber-incident is identified and reported. Detecting most incidents requires automated analysis tools, system behavior patterns, and an awareness of what to look for among operators, supervisors, and other staff. The building equipment operators and the process engineers are usually critical to detection of unusual operations and are the first to note a difference in system behavior. Also called *discovery*.

incident description: Clear definitions of what is a security incident must be identified and communicated to the extent possible. This is particularly important when considering if equipment failure or unexpected software behavior is caused by a cybersecurity incident, due to mechanical failure because of wear, environmental conditions, or other non-security related factors. It is important to understand and differentiate between a cybersecurity and non-cybersecurity incident. Many IT-type incidents are fairly easily classified. These include denial-of-service attacks, unauthorized access to networks, accessing protected and private information, defacing web pages, misuse of services, and so forth.

incident handling: An adverse event in a network or a building control system or the threat of the occurrence of such an event. SOURCE: SP 800-61

incident response plan: The documentation of a predetermined set of instructions or procedures to detect, respond to, and limit consequences of a malicious cyber-attack against an organization's information system(s). SOURCE: SP 800-34

incomplete parameter checking: System flaw that exists when the operating system does not check all parameters fully for accuracy and consistency, thus making the system vulnerable to penetration. SOURCE: CNSSI-4009

incremental encoder: Encoders are typically used in positioning and motor speed feedback applications. They provide change information only, so require a reference device to calculate motion.

inculpatory evidence: Evidence that tends to increase the likelihood of fault or guilt. SOURCE: SP 800-72

indicator: Recognized action, specific, generalized, or theoretical, that an adversary might be expected to take in preparation for an attack. SOURCE: CNSSI-4009

indirect attack: (1) A cyber-attack launched from a third party computer making it difficult to track the origin. 2) When a hacker cannot penetrate a building's defenses directly, he can resort to an indirect cyber-physical attack by disrupting vital utilities outside the facility such as natural gas, power, water and sanitary sewage system. Shutting down any one of these would have a second order effect on a building rendering the facility unusable.

induced noise: Noise that obscures an electrical signal.

Industrial Control System (ICS): An information system used to control industrial processes such as manufacturing, product handling, production, and distribution. Industrial control systems include supervisory control and data acquisition systems (SCADA) used to control geographically dispersed assets, as well as distributed control systems (DCS) and smaller control systems using programmable logic controllers to control localized processes. SOURCE: SP 800-53; SP 800-53A

Inetd (xinetd): An Internet service daemon that listens for request from other Internet services like FTP, Telnet, and POP.

inference attack: A data mining technique performed by analyzing data in order to illegitimately gain knowledge about a subject or database. Sensitive information can be considered leaked if an adversary can infer its real value with a high degree of confidence.

information assurance (IA): Measures that protect and defend information and information systems by ensuring their availability, integrity, authentication, confidentiality, and non-repudiation. These measures include providing for restoration of information systems by incorporating protection, detection, and reaction capabilities. SOURCE: SP 800-59; CNSSI-4009

information assurance component (IAC): An application (hardware and/or software) that provides one or more information assurance capabilities in support of the overall security and operational objectives of a system. SOURCE: CNSSI-4009

information disclosure: A breach that occurs when information that is thought to be secured is released to uncleared individuals.

information domain: A three-part concept for information sharing, independent of, and across information systems and security domains that (1) identifies information sharing participants as individual members, (2) contains shared information objects, and (3) provides a security policy that identifies the roles and privileges of the members and the protections required for the information objects. SOURCE: CNSSI-4009

information warfare: (1) The competition between defensive and offensive adversaries over information resources. (2) Dan Kuehl: "conflict or struggle between two or more groups in the information environment."

information security: The protection of information and information systems from unauthorized access, use, disclosure, disruption, modification, or destruction in order to provide confidentiality, integrity, and availability. SOURCE: SP 800-37; SP 800-53; SP 800-53A; SP 800-18; SP 800-60; CNSSI-4009; FIPS 200; FIPS 199; 44 U.S.C., Sec. 3542

information security architecture: An embedded, integral part of the enterprise architecture that describes the structure and behavior for an enterprise's security processes, information security systems, personnel and organizational subunits, showing their alignment with the enterprise's mission and strategic plans. SOURCE: SP 800-39

Information Security Continuous Monitoring (ISCM): Maintaining ongoing awareness of information security, vulnerabilities, and threats to support organizational risk management decisions. The terms *continuous* and *ongoing* in this context mean that security controls and organizational risks are assessed and analyzed at a frequency sufficient to support risk-based security decisions to adequately protect organization information. SOURCE: SP 800-137

Information Security Continuous Monitoring (ISCM) process: A process to do the following:

- Define an ISCM strategy
- Establish an ISCM program
- Implement an ISCM program
- Analyze data and Report findings
- Respond to findings
- Review and update the ISCM strategy and program

SOURCE: SP 800-137

information system: A discrete set of information resources organized for the collection, processing, maintenance, use, sharing, dissemination, or disposition of information. SOURCE: FIPS 200; FIPS 199; SP 800-53A; SP 800-37; SP 800-60; SP 800-18; 44 U.S.C., Sec. 3502; OMB Circular A-130, App. III

information technology system: An information system that is automated or is an assembly of computer hardware and software configured for the purpose of classifying, sorting, calculating, computing, summarizing, transmitting and receiving, storing and retrieving data with a minimum of human intervention. The term includes single application programs, which operate independently of other program applications. A *sensitive information technology system* means an information technology system that contains sensitive information. SOURCE: SP 800-53; SP 800-53A; SP 800-37; SP 800-18; SP 800-60; FIPS 200; FIPS 199; CNSSI-4009; 40 U.S.C., Sec. 11101 and Sec 1401

information system resilience: The ability of a building control system to (1) continue to operate under stress, even if in a degraded condition, while maintaining essential building operational capabilities; and (2) recover effectively and quickly. SOURCE: SP 800-30

infrastructure: (1) The basic physical structures, facilities and utilities (e.g., roads, buildings, and water and power distribution systems) needed by a society. (2) The physical equipment (computers, cases, racks, cabling, etc.) that comprises a building control system. (3) the foundational basis that supports the building control system capabilities, including the telecommunications and network connectivity.

ingress filtering: The process of blocking incoming packets that use obviously false IP addresses, such as reserved source addresses. SOURCE: SP 800-61

initiator: A Small Computer System Interface (SCSI) device that requests another SCSI device (a target) to perform an operation; usually a host computer acts as an initiator and a peripheral device acts as a target.

InputBuffer: Refers to computer memory, used to store data as they are received before it continues to the CPU for processing.

Input/Output (I/O): A general term for the method that is used to communicate with a computer as well as the data involved in the communications. Read or write computer signals; the connection path between the CPU bus and the hard drives.

input validation attack: A cyber-attack when a hacker sends unexpected input to a server in the hopes of confusing a building controls system.

insider attack: An entity inside the security perimeter that is authorized to access building control system resources, but uses them in a way not approved by those who granted the authorization. In addition to intentional violators, other types of insiders are:

- **Exploited insiders** may be "tricked" by external parties into providing data or passwords they shouldn't.

- **Careless insiders** may simply press the wrong key and accidentally delete or modify critical information.

- **Malicious insiders** intend to cause mischief, such as a disgruntled employee.

insider threat: An entity with authorized access that has the potential to harm an information system through destruction, disclosure, modification of data, and/or denial-of-service. SOURCE: SP 800-32

inspectable space: Three dimensional space surrounding equipment that processes classified and/or sensitive information within which TEMPEST exploitation is not considered practical or where legal authority to identify and remove a potential TEMPEST exploitation exists. Synonymous with *zone of control*. SOURCE: CNSSI-4009

interrupt: A signal that informs the operating system that something that something needs immediate attention. The interrupt informs the processor to a priority condition.

Institute of Electrical and Electronics Engineers (IEEE): The largest technical society in the world, consisting of engineers, scientists, and students; has declared standards for computers and communications.

Integrated Workplace Management System (IWMS): A software platform that helps organizations optimize the use of workplace resources, including the management of a company's real estate portfolio, infrastructure, facilities, project management, energy use, and sustainability. SOURCE: Wikipedia

Integrity, Data: That attribute of data relating to the preservation of (1) its meaning and completeness, (2) the consistency of its representation(s), and (3) its correspondence to what it represents. The building control system or installation contains information that must be protected from unauthorized, unanticipated, or unintentional modification or destruction, including detection of such activities. Integrity is important to all information because inaccuracy compromises the value of the building control system. SOURCE: SP 800-53; SP 800-53A; SP 800-18; SP 800-27; SP 800-37; SP 800-60; FIPS 200; FIPS 199; 44 U.S.C., Sec. 3542

Integrity, System: That attribute of a building control system when it performs its intended function in an unimpaired manner, free from deliberate or inadvertent unauthorized manipulation of the building control system. SOURCE: CNSSI-4009

intellectual property theft: A theft of copyrighted material where it violates copyrights and patents. A cybercrime intended to steal trade secrets, patents and research. Theft of an idea, plan, or methodology. SOURCE: SP 800-32

intelligent electronic device (IED): A device capable of two-way communication directly with a BCS, ICS, or SCADA computer that performs electrical functions such as sensors, actuators, servos, relays and circuit breakers. Any device incorporating one or more processors with the capability to receive or send data/control from or to an external source (e.g., electronic multifunction meters, digital relays, controllers).

Inter-Control Center Communications Protocol (ICCP): The SCADA protocol used to exchange information with business partners or to exchange information between the corporate network and control center network.

interlock: An electrical or mechanical device used to prevent a machine from harming its operator or damaging itself by preventing one element from changing state due to the state of another element, and vice versa. A simple example is the elevator interlock that prevents a moving elevator from opening its doors, and prevents the stationary elevator (with open doors) from moving.

Internet Control Message Protocol (ICMP): One of the main Internet protocols. Used by network devices to send error messages such as when a requested service is not available.

interface: Common boundary between independent systems or modules where interactions take place. SOURCE: CNSSI-4009

interface control document: Technical document describing interface controls and identifying the authorities and responsibilities for ensuring the operation of such controls. This document is baselined during the preliminary design review and is maintained throughout the information system life cycle. SOURCE: CNSSI-4009

InterFacility Contingency Planning Regulation: A regulation written and imposed by the Federal Financial Institutions Examination Council concerning the need for financial institutions to maintain a working disaster recovery plan. SOURCE: FFIEC

interim organizational structure: An alternate organization structure that will be used during recovery from a cyber-attack. This temporary structure will typically streamline the chain of command and increase decision-making autonomy.

internal hot sites: A fully equipped alternate data processing site.

Internet: The Internet is the single, interconnected, worldwide system of commercial, governmental, educational, and other computer networks that share (a) the protocol suite specified by the Internet Architecture Board (IAB), and (b) the name and address spaces managed by the Internet Corporation for Assigned Names and Numbers (ICANN). SOURCE: CNSSI-4009

Internet appliance: An Internet appliance (sometimes called a *Net appliance*, a *smart appliance* or an *information appliance*) is a machine designed for a specific function that also has a built-in web-enabled computer.

Internet-based Mobile Ad hoc Networks (iMANETs): Ad hoc networks that link mobile nodes and fixed Internet-gateway nodes.

Internet of Things (IoT): The network of physical objects or "things" embedded with electronics, software, sensors, and connectivity to enable devices to exchange data with other connected devices. The down side to all this is that the IoT can become a cheap tool for remote surveillance and reconnaissance and a great deal of information can be learned from device usage behavior. By aggregating this surveillance information about your building over time, an attacker could get a very accurate understanding of your building operations.

Internet tools such as Wink allow a person to control, from a single screen, his Internet-connected home devices, such as door locks, window shades, and LED lights. Anyone capable of hacking your Wink account may be able to identify your social media accounts, the names of your devices (like Lou's iPad) and your network information. An app that monitors your grill's propane tank may record the tank's latitude and longitude, thus revealing the exact location of your house. Hacking into a Nest thermostat would allow someone to figure out when your house was occupied and when it was not. Manufacturers of IP-enabled devices say you can opt out of sharing information with vendors, software developers and third-party applications, but you may not be aware how much information their device is collecting. SOURCE: Wikipedia

Internet Protocol (IP): A data-oriented protocol used for communicating data across a packet-switched inter-network. Data is sent in blocks referred to as *packets*. Standard protocol for transmission of data from source to destinations in packet-switched communications networks and interconnected systems of such networks. SOURCE: CNSSI-4009

Internet Protocol Security (IPsec): Cryptographic protocols for securing packet flows and key exchange.

Internet Service Provider (ISP): A company that provides Internet access services to consumers and businesses; ISPs lease connections from Internet backbone providers. Most ISPs are small companies that service a local area; however, there are also regional and national ISPs (such as America Online).

interoperability: The ability of one building control system to control another, even though the two building control systems are made by different manufacturers.

interruption: An outage caused by the failure of communications links outside of the local facility.

intranet: A private network that is employed within the confines of a given enterprise (e.g., internal to a business or agency). SOURCE: CNSSI-4009

intruder: A person who is the perpetrator of a computer security incident. Intruders are often referred to as *hackers* or *crackers*. While hackers were very technical experts in the early days of computing, this term was later used by the media to refer to people who break into other building control systems. Crackers is based on hackers and the fact that these people "crack" computer systems and security barriers. Most of the time *cracker* is used to refer to more notorious intruders and computer criminals. Sometimes it is argued that the term *attacker* would be better, as an unsuccessful attack doesn't constitute an intrusion. But, because of the intention of the person responsible for the attack, the term *intruder* is used. SOURCE: 800-61

intrusion: An unauthorized act of bypassing the security mechanisms of a network or building control system. SOURCE: CNSSI-4009

intrusion detection system (IDS): Hardware or software product that gathers and analyzes information from various areas within a computer or a network to identify possible security breaches, which include both intrusions (attacks from outside the organizations) and misuse (attacks from within the organizations) SOURCE: CNSSI-4009

intrusion detection systems (IDS):

- **Host-based** IDS operate on information collected from within an individual computer system. This vantage point allows host-based IDSs to determine exactly which processes and user accounts are involved in a particular attack on the Operating System. Furthermore, unlike network-based IDSs, host-based IDSs can more readily "see" the intended outcome of an attempted attack, because they can directly access and monitor the data files and system processes usually targeted by attacks. SOURCE: SP 800-36; CNSSI-4009

- **Network-based** IDS detect attacks by capturing and analyzing network packets. Listening on a network segment or switch, one network-based IDS can monitor the network traffic affecting multiple hosts that are connected to the network segment. SOURCE: SP 800-36; CNSSI-4009

intrusion prevention system (IPS): System(s) detect an intrusive activity and can attempt to stop the activity, ideally before it reaches its targets. SOURCE: SP 800-36; CNSSI-4009

investigation: A systematic and formal inquiry into a qualified threat or incident using digital forensics and perhaps other traditional criminal inquiry techniques to determine the events that transpired and to collect evidence. SOURCE: NICCS

I/O brick: A PLC storage system that handles a huge amount of I/O requests from machines and desktops.

IP camera cyber-attack: Hack into a BCS through an IP camera as the attack vector.

IP-controlled device: An intelligent electronic device that can be controlled over the Internet. So, an IP controlled rack-mount power controller allows remote access, real-time monitoring and customer management from a phone, computer, or tablet. Such a device would be useful to reboot a server, but this represents a huge security risk. I don't recommend these if you want a secure building.

IP-enabled door lock: A door lock that allows a user to control and manage building access with a smartphone or tablet using a web app. Some *smart locks* notify when certain people enter or leave, and they can be customized for exactly how long a particular individual will have access, what days or even between what hours. Some smart locks have a built-in camera that takes a photo of the person who activates the lock. The security issues are too numerous to mention here.

IP forwarding: Also known as Internet *routing* is a process used to determine which path a packet or datagram can be sent. The process uses routing information to make decisions and is designed to send a packet over multiple networks. Generally, networks are separated from each other by routers.

IP flood attack: A denial-of-service attack that sends a host more "ping" packets than the protocol can handle.

IP masquerading (IPMASQ): Network address translation (NAT) that allows internal computers that don't have an officially assigned IP address to communicate to other networks and the Internet. It allows one machine to act on behalf of other machines. Also called **MASQ.**

I/Os per second (IOPS): A measure of performance for a host-attached storage device or RAID controller.

islanding: Islanding refers to the condition in which a *distributed generator (DG)* continues to power a location even though electrical grid power from the electric utility is no longer present. Islanding can be dangerous to utility workers, who may not realize that a circuit is still powered, and it may prevent automatic reconnection of devices. For that reason, distributed generators must detect islanding and immediately stop producing power the commercial power is restored; this is referred to as *anti-islanding.*

isolation valve: A valve in a fluid handling system that stops the flow of process media to a given location, usually for maintenance or safety purposes.

IT manager: The individual responsible for the information system infrastructure related to the ICS. This includes enclave perimeter devices, network backbone, servers, and workstations.

IT-related risk: The net mission/business impact considering

- the likelihood that a particular threat source will exploit, or trigger, a particular information system vulnerability, and

- the resulting impact if this should occur. IT-related risks arise from legal liability or mission/business loss due to, but not limited to:

  a. Unauthorized (malicious, non-malicious, or accidental) disclosure, modification, or destruction of information;

  b. Non-malicious errors and omissions;

  c. IT disruptions due to natural or man-made disasters; or

  d. Failure to exercise due care and diligence in the implementation and operation of the IT.

SOURCE: SP 800-27

IT security architecture: A description of security principles and an overall approach for complying with the principles that drive the system design; that is, guidelines on the placement and implementation of specific security services within various distributed computing environments. SOURCE: SP 800-27

CHAPTER 11

■ ■ ■

J

jailbreaking: The process of removing hardware restrictions on iOS, Apple's operating system, on devices running it through the use of software exploits; such devices include the iPhone, iPod touch, iPad, and second-generation Apple TV. Jailbreaking permits root access to the iOS file system and manager, allowing the download of additional applications. Jailbreaking is a form of privilege escalation. The name refers to breaking the device out of its "jail," which is a technical term used in Unix-style systems, for example in the term *FreeBSD jail*. Computer criminals may jailbreak an iPhone to install malware, or target jailbroken iPhones on which malware can be installed more easily. Jailbreaking is legal in the United States, but may not be in other countries. The following are the types of jailbreaking:

- An **untethered** jailbreak has the property that if the user turns the device off and back on, the device will start up completely, and the kernel will be patched without the help of a computer: thus enabling the user to boot without the need to use a computer.

- With a **tethered** jailbreak, a computer is needed to turn the device on each time it is rebooted. If the device starts back up on its own, it will no longer have a patched kernel, and it may get stuck in a partially started state.

- A **semi-tethered** jailbreak means that when the device boots, it will no longer have a patched kernel (so it will not be able to run modified code), but it will still be usable for normal functions such as making phone calls, or texting.

jamming attack: An attack in which a device is used to emit electromagnetic energy on a wireless network's frequency to make it unusable. SOURCE: SP 800-48

jitter: The time or phase difference between the data signal and the ideal clock.

jockey pump: A small multistage centrifugal pump connected to a fire sprinkler system designed to maintain pressure in a fire protection piping system to an artificially high level so that the operation of a single fire sprinkler caused a pressure drop that is sensed by the fire pump automatic controller, causing the fire pump to start. The jockey pump is essentially a portion of the fire pump's control system; hence a jockey pump is an important part of the fire pump control system. SOURCE: Wikipedia

Joint Personnel Adjudication System (JPAS): Standardizes the adjudication process within DoD, and provides a database and processes within the security manager realm of functions. SOURCE: DSS FCL Orientation Handbook

jump kit: A container that has the items necessary to respond to a cyber-physical attack to help mitigate the effects of delayed reactions. It contains the tools the Incident Response Teams will need to restore a system to its last fully mission-capable state. Knowing what the Recovery point should be is the key to ensuring all known remnants of an attack have been removed from all components of the BCS. This means all hardware and software are configured in accordance with operational requirements, and checksums and hashes are in conformance with vendor specifications.

jump-kit contents:

- Incident Notifications List: document contact information for the information assurance manager

- Document stakeholders who could be affected by a cyber-physical attack on BCS

- Established notification procedures with management or chain of command

- Universal serial bus (USB) drives, bootable USB (or LiveCD) with up-to-date anti-malware, and other software tools that can read and/or write to file system

- Laptop with anti-malware utilities and Internet access (for downloads)

- Computer and network tool kit to add components, hard drives, connectors, wire cables, etc.

- Hard disk duplicators with write-block capabilities to capture hard drive images

- Firewall access control lists

- Firewall hard disk image

- IDS rules

- IDS image

- Back up of firewall, router, and switch IOS

- Backup of PLC configurations and firmware

- Backup RTU software, database, and configurations

- Back up of all other computer assets to include HMI, Historian, and Database

- Network map of all expected connections to the BCS

just a bunch of disks (JBOD): A hard drive array without a built-in controller.

CHAPTER 12

K

kerberos: A system developed at MIT that depends on passwords and symmetric cryptography to implement ticket-based, peer entity authentication service and access control service distributed in a client-server network environment. SOURCE: SP 800-63

kernel: The core of an operating system such as Windows 98, Windows NT, Mac OS, or Unix; provides basic services for the other parts of the operating system, making it possible for it to run several programs at once (multitasking), read and write files and connect to networks and peripherals.

kewl: Hacker-speak for "cool," normally used ironically. When crackers praise something they use the prefix *uber* (from German).

key bundle: The three cryptographic keys (Key1, Key2, Key3) that are used with a Triple Data Encryption Algorithm mode. SOURCE: SP 800-67

key expansion: Routine used to generate a series of Round Keys from the Cipher Key. SOURCE: FIPS 197

key loader: A self-contained unit that is capable of storing at least one plaintext or encrypted cryptographic key or key component that can be transferred, upon request, into a cryptographic module. SOURCE: FIPS 140-2

key management personnel (KMP): Individuals required to be cleared or excluded as determined by the Cognizant Security Agency or when delegated, its CSO, as part of issuing a Facility Security Clearance. SOURCE: DSS FCL Orientation Handbook

key stream: Sequence of symbols (or their electrical or mechanical equivalents) produced in a machine or auto-manual cryptosystem to combine with plain text to produce cipher text, control transmission security processes, or produce key. SOURCE: CNSSI-4009

key stretching: Passwords or *passphrases* created by humans are often short or predictable enough to allow password cracking. Key stretching makes such attacks more difficult by using cryptography techniques to make a weak password more secure against a brute force attack by increasing the time it takes to test each possible key. The initial key is fed into an algorithm that outputs an *enhanced* key (e.g., at least 128 bits).

keystroke logger attack: A program or USB device designed to record which keys are pressed on a computer keyboard used to obtain passwords or encryption keys and thus bypass other security measures. Another type of keystroke logger uses the accelerometer in a smartphone to capture keystrokes. SOURCE: SP 800-82

keystroke monitoring: The process used to view or record both the keystrokes entered by a computer user and the computer's response during an interactive session. Keystroke monitoring is usually considered a special case of audit trails. SOURCE: SP 800-12; CNSSI-4009

© Luis Ayala 2016
L. Ayala, *Cybersecurity Lexicon*, DOI 10.1007/978-1-4842-2068-9_12

kill chain: Kill chain analysis illustrates the sequence of stages that the hacker must progress through successfully before achieving the desired objective; just one mitigation can disrupt the chain and stop the cyber-attack. The stages are: reconnaissance and staging, delivery and attack, exploitation and installation, system compromise. A kill chain analysis describes the very structure of a cyber-intrusion and the corresponding model guides analysis to inform actionable cyber security intelligence. Through this model, defenders can develop resilient mitigations against intruders and prioritize investments in new defensive technology or processes. SOURCE: Lockheed Martin

kill switch: A kill switch, also known as an *emergency stop, e-stop*, and *Big Red Button*, is a safety mechanism used to shut off a device in an emergency situation in which it cannot be shut down in the usual manner. Unlike a normal shut-down switch/procedure, which shuts down all building control systems in an orderly fashion and turns the machine off without damaging it, a kill switch is designed and configured to (1) completely and as quickly as possible abort the operation, even if this damages equipment and (2) be operable in a manner that is quick, simple (so that even a panicking operator with impaired executive function can activate it), and, usually, (3) be obvious even to an untrained operator or a bystander. In a data center it is called a scram switch.

kinetic cyber-attack: A cyber-physical attack that is intended to cause physical damage in the real world to people, buildings, equipment, infrastructure or a nation's way of life. Not a virtual attack or theft of data.

CHAPTER 13

L

laboratory attack: Use of sophisticated signal recovery equipment in a laboratory environment to recover information from data storage media. SOURCE: SP 800-88; CNSSI-4009

ladder diagram: A number of control subsets provide building equipment operating limits and safety controls. A ladder diagram indicates sensing devices specifically designed to prevent the operation of the unit unless a preset number of conditions are met. The circuits are wired in series, which means all conditions must be satisfied prior to allowing the unit to operate. For chillers, some of the more common items are high head pressure, low suction temperature, proof of water flow at the condenser and evaporator, oil pressure/temperature, starting current, and voltage.

LADDERLOGIX: Programming language that represents a program by a graphical diagram based on the circuit diagrams of relay logic hardware.

lamer: Cracker-speak to describe someone who does not really understand what he is doing.

LAND attack: LAND (Local Area Network Denial) attack is a DoS (denial-of-service) attack that consists of sending a special poison spoofed packet to a computer, causing it to lock up. The attack involves sending a spoofed TCP SYN packet (connection initiation) with the target host's IP address to an open port as both source and destination. This causes the machine to reply to itself continuously. Most firewalls should intercept and discard the poison packet thus protecting the host from this attack. Some operating systems released updates fixing this security hole. In addition, routers should be configured with both ingress and egress filters to block all traffic destined for a destination in the source's address space, which would include packets where the source and destination IP addresses are the same.

LAN recovery: The component of BCS recovery which deals specifically with the replacement of LAN equipment in the event of a disaster, and the restoration of essential data and software.

laptop Webcam Hack: Most laptops are equipped with a built-in webcam and microphone and both can be hacked. An attacker can listen in or watch you work at your laptop without your knowledge. This is a fairly simple hack made possible by Trojan horse malware called *Blackshades* that even a script kiddie can master.

leak alarm: Used to detect and report water leaks.

leak before burst: A pressure vessel design such that a crack in the vessel will grow through the wall, allowing the contained fluid to escape (reducing the pressure) prior to growing so large as to cause fracture at the operating pressure.

leased line: Synonymous with a dedicated line. The telecommunications line between two locations that provided in exchange for a monthly rent.

least privilege: The security objective of granting users only those accesses they need to perform their official duties. SOURCE: SP 800-12

© Luis Ayala 2016
L. Ayala, *Cybersecurity Lexicon*, DOI 10.1007/978-1-4842-2068-9_13

leech: Among BBS types, crackers and warez d00dz, one who consumes knowledge without generating new software, cracks, or techniques. Someone that downloads files with few uploads. Also known as a *troughie* (a pig at a trough).

legacy: A computer, building control system, or software that was created for a specific purpose, but is now outdated; anything left over from a previous version of the BCS hardware or software.

level of concern: Rating assigned to an information system indicating the extent to which protection measures, techniques, and procedures must be applied. High, Medium, and Basic are identified levels of concern. A separate Level-of-Concern is assigned to each information system for confidentiality, integrity, and availability. SOURCE: CNSSI-4009

level of protection: Extent to which protective measures, techniques, and procedures must be applied to building control systems and networks based on risk, threat, vulnerability, building control system interconnectivity considerations, and information assurance needs. SOURCE: CNSSI-4009

Levels of protection are:

- **Basic**: building control systems and networks requiring implementation of standard minimum security countermeasures.

- **Medium**: building control systems and networks requiring layering of additional safeguards above the standard minimum security countermeasures.

- **High**: building control systems and networks requiring the most stringent protection and rigorous security countermeasures.

life-critical system: A building control system whose failure or malfunction may result in one (or more) of the following outcomes: death or serious injury to people, loss or severe damage to equipment or property, or environmental harm. Several reliability regimes for life-critical building control systems exist:

- **Fail-operational** building control systems continue to operate when their control systems fail.

- **Fail-safe** building control systems become safe when they cannot operate.

- **Fail-secure** building control systems maintain maximum security when they cannot operate.

- **Fail-passive** building control systems continue to operate in the event of a system failure.

- **Fault-tolerant** building control systems avoid service failure when faults are introduced to the system.

light fidelity (Li-Fi): Wireless data streaming using LED lights to transmit information. LEDs can communicate twice as fast (15 gigabits per second) as Wi-Fi. Li-Fi may be more secure because light can't go through walls, hackers would not be able to log on to Li-Fi networks in the same way that they're able to eavesdrop on Wi-Fi. Li-Fi may also be *less* secure.

lighting controls: Provide comprehensive control of the whole building lighting system using energy management strategies for increased energy savings.

lighting optimization: Lighting levels in US buildings typically exceed necessary illumination levels, resulting in reduced energy efficiency. Lighting optimization techniques such as automated lighting schedules and occupancy or vacancy sensors ensure lighting is adequate for the spaces when needed based on the layout, without over-illuminating the space and wasting energy. Task lighting such as desk lamps with an occupancy or vacancy sensor can be another effective way to reduce overall office lighting levels.

light tower: A device containing a series of indicator lights and an embedded controller used to indicate the state of a building control system component based on an input signal.

like-jacking cyber-attack: When criminals post fake Facebook Like buttons to web pages. Users who click the button don't "like" the page, but instead download malware. SOURCE: FBI Internet Social Networking Risks

likelihood of occurrence: In information assurance risk analysis, a weighted factor based on a subjective analysis of the probability that a given threat is capable of exploiting a given vulnerability. SOURCE: CNSSI-4009

limited maintenance: COMSEC maintenance restricted to fault isolation, removal, and replacement of plug-in assemblies. Soldering or unsoldering usually is prohibited in limited maintenance. SOURCE: CNSSI-4009

line conditioning: Elimination of unintentional signals or noise induced or conducted on a telecommunications or building control system signal, power, control, indicator, or other external interface line. SOURCE: CNSSI-4009

line conduction: Unintentional signals or noise induced or conducted on a telecommunications or building control system signal, power, control, indicator, or other external interface line. SOURCE: CNSSI-4009

line rerouting: A service offered by many regional telephone companies allowing a data center to quickly reroute the network of dedicated lines to a backup site.

line voltage regulators: Also known as *surge protectors*. These protectors/regulators distribute electricity evenly.

linear tape open (LTO): A standard tape format developed by HP, IBM, and Seagate.

lineup: A complete assembly of electrical switchgear. The term entrance applies to the incoming bay or device, sometimes referred to as a *main. Load-feeder* applies to the bay or device for an outgoing circuit, sometimes referred to as a *tap*. The term *bay* is used in switch/fuse gear and it is a single unit that is the equivalent of a vertical section in circuit-breaker gear. SOURCE: S&C Electric Company

link farm: A group of web sites that all hyperlink to every other site in the group. A link farm is a form of spamming the index of a search engine (sometimes called *spamdexing*). Although some link farms can be created by hand, most are created through automated programs.

link-jacking cyber-attack: Used to redirect a trusted web site's links to a malware-infected web site that hides drive-by downloads or other types of infections.

Link Service Access Point (LSAP): An LSAP is a 3-byte header including a destination, a source field, and a control field. A service access point is a label for network endpoints in OSI networks.

Link Service Data Unit (LSDU): User data received from the network layer.

LNS plug-in: Software that runs in a network configuration tool. Configuration plug-ins provide a "user-friendly" method to edit a device's configuration properties.

load shed: The deliberate shutdown of electric power in a part or parts of a power-distribution system, generally to prevent the failure of the entire electric system when the demand strains the capacity of the system.

local access: Access to an organizational information system by a user (or process acting on behalf of a user) communicating through a direct connection without the use of a network. SOURCE: SP 800-53; CNSSI-4009

local area network (LAN): Computers connected together so that they can communicate with each other. A network of computers, within a limited area (e.g., a company or organization); computing equipment, in close proximity to each other, connected to a server which houses software that can be accessed by the users. This method does not utilize a public carrier.

local mode: Equipment is operated manually through a keypad located at the device, independent of the BCS. Also referred to as *hand mode* or *local control.*

lock bypass: A lock bypass is a technique in lock-picking, of defeating a lock through unlatching the underlying locking mechanism without operating the lock at all. Locks may be bypassed by a variety of techniques including loiding; that is, the "credit card" technique.

Lockout-Tagout (LOTO): Lock and tag is a safety procedure, which is used in industry and research settings to ensure that dangerous machines are properly shut off and not able to be started up again prior to the completion of maintenance or servicing work. It requires that hazardous energy sources be "isolated and rendered inoperative" before work is started on the equipment in question. The isolated power sources are then locked and a tag is placed on the lock identifying the worker who has placed it. The worker then holds the key for the lock ensuring that only he or she can start the machine. This prevents accidental startup of a machine while it is in a hazardous state or while a worker is in direct contact with it. Lockout-tagout is used across industries as a safe method of working on hazardous equipment and is mandated by law in some countries. SOURCE: Wikipedia

log clipping: Selective removal of building control system log entries to hide a compromise.

logic bomb attack: A piece of code intentionally inserted into a software system that will set off a malicious function when specified conditions are met. SOURCE: CNSSI-4009

logical completeness measure: Means for assessing the effectiveness and degree to which a set of security and access control mechanisms meets security specifications. SOURCE: CNSSI-4009

logical unit number (LUN): An addressing scheme used to define SCSI devices on a single SCSI bus.

Long-Term Exposure Limit (LTEL): The eight-hour LTEL is the time-weighted average concentration for a normal 8-hour day to which most workers may be repeatedly exposed, day after day, without adverse effect.

LonMark Object: A collection of network variables, configuration properties, and associated behavior and described by a Functional Profile that defines how information is exchanged between devices (developed by LonMark International).

LonMark: A certification issued by LonMark International.

LonMark International: Standards committee consisting of numerous independent product developers, building control system integrators and end users dedicated to determining and maintaining the interoperability guidelines for LonWorks.

LonWorks: The term used to refer to the overall technology related to the CEA-709.1-C protocol (sometimes called LonTalk), including the protocol itself, network management, interoperability guidelines and products.

LonWorks Network Services (LNS): A network management and database standard for CEA-709.1-C devices.

loopback address (127.0.0.1): Pseudo IP address that always refers back to the local host and never sent out onto a network.

loss: The unrecoverable building resources that are damaged as a result of a cyber-physical attack. Such losses may be loss of life, revenue, market share, competitive stature, public image, facilities, or operational capability.

loss reduction: The technique of instituting mechanisms to lessen the exposure to a particular risk. Loss reduction is intended to react to an event and limit its effect. An example of loss reduction is a sprinkler system.

low impact: The loss of confidentiality, integrity, or availability that could be expected to have a limited adverse effect on organizational operations, organizational assets, individuals, other organizations, or the national security interests of the United States. (1) causes a degradation in mission capability to an extent and duration that the organization is able to perform its primary functions, but the effectiveness of the functions is noticeably reduced; (2) results in minor damage to organizational assets; (3) results in minor financial loss; or (4) results in minor harm to individuals). SOURCE: CNSSI-4009

low-level alarm: Visual and audible alarms when the liquid level drops below a preset level.

lower explosive limit (LEL): The lowest concentration of "fuel" in the air that will burn. For most flammable gases and vapors, it is less than 5% by volume. The LEL% is the percentage of the lower explosive limit (for example, 10% LEL of methane is approx. 0.5% by volume).

low probability of detection: Result of measures used to hide or disguise intentional electromagnetic transmissions. SOURCE: CNSSI-4009

low probability of intercept: Result of measures to prevent the intercept of intentional electromagnetic transmissions. The objective is to minimize an adversary's capability of receiving, processing, or replaying an electronic signal. SOURCE: CNSSI-4009

Luminaire Control Module (LCM): Provides an interface between light fixture ballast and the communications network.

luser: In Internet slang, a luser is an annoying, stupid or irritating computer user. The word is a blend of *loser* and *user*. Among hackers, the word luser takes on a broad meaning, referring to any normal user (in other words, not a "guru"), with the implication the person is also a loser. The term is interchangeable with the hacker term *lamer*.

luser attitude readjustment tool (LART): Also known as a *clue-by-four*, *cluestick*, or *cluebat*, meaning turning off the user's access to computer resources.

CHAPTER 14

M

machine controller: A control system/motion network that electronically synchronizes drives within a machine system instead of relying on synchronization via mechanical linkage.

machine learning & evolution: A field concerned with designing and developing artificial intelligence algorithms for automated knowledge discovery and innovation by computer systems.

machine-readable media: Media that can convey data to a given sensing device; for example, diskettes, disks, tapes, and computer memory.

machine vision (MV): The technology and methods used to provide imaging-based automatic inspection and analysis for such applications as automatic inspection, process control, and robot guidance in industry.

macro virus: A type of malicious code that attaches itself to documents and uses the macro programming capabilities of the document's application to execute, replicate, and spread or propagate itself. SOURCE: SP 800-61; CNSSI-4009

magnetic ink character reader (MICR) equipment: Equipment used to imprint machine-readable code. Financial institutions use this to prepare paper data for processing, encoding (imprinting) items such as routing and transit numbers, account numbers, and dollar amounts.

magnetic remanence: Magnetic representation of residual information remaining on a magnetic medium after the medium has been cleared.

mainframe computer: A high-end computer processor, with related peripheral devices, capable of supporting large volumes of batch processing, high performance on-line transaction processing systems, and extensive data storage and retrieval. Similar Term: *Host Computer.*

maintenance hook: Special instructions (trapdoors) in software allowing easy maintenance and additional feature development. Since maintenance hooks frequently allow entry into the code without the usual checks, they are a serious security risk if they are not removed prior to live implementation. SOURCE: CNSSI-4009

makeup water: Water fed to a system to replace what is lost; for example, water fed to a boiler to replace what's lost as steam or condensate; water fed to a cooling tower to replace that lost by evaporation, drift, or other causes.

malicious applet attack: A small application program that is automatically downloaded and executed and that performs an unauthorized function on a building control system. SOURCE: CNSSI-4009

malicious code attack: Program code intended to perform an unauthorized function or process that will have adverse impact on the confidentiality, integrity, or availability of a building control system. SOURCE: SP 800-53; CNSSI-4009

© Luis Ayala 2016

L. Ayala, *Cybersecurity Lexicon*, DOI 10.1007/978-1-4842-2068-9_14

malicious command and control: A method for unauthorized remote identification of, access to, or use of, an information system or information that is stored on, processed by, or transiting an information system.

malicious logic attack: Hardware, firmware, or software that is intentionally included or inserted in a building control system to perform an unauthorized function or process that will have adverse impact on the confidentiality, integrity, or availability of a building control system. SOURCE: CNSSI-4009

malicious reconnaissance: A method for actively probing or passively monitoring an information system for the purpose of discerning security vulnerabilities of the information system, if such method is associated with a known or suspected cybersecurity threat.

malicious software attack: Any of a family of computer programs developed with the sole purpose of doing damage. Malicious code is usually embedded in software programs that appear to provide useful functions but, when activated by a user, cause undesirable results.

malvertising attack: A method whereby users download malicious code by simply clicking on a malicious advertisement placed on a web site by cyber-criminals without the knowledge of the web site owner. The malverts easily pass off as genuine advertisements on a web site that is infected.

malware attack: Malicious software designed to infiltrate or damage a building control system. Software or firmware intended to perform an unauthorized process that will have adverse impact on the confidentiality, integrity, or availability of a building control system. Malware types include virus, worm, Trojan horse, root kit, spyware and adware designed to infect a host. Spyware and some forms of adware are also examples of malicious code (malware). SOURCE: SP 800-83

man-in-the-middle (MITM) attack: A cyber-attack where the attacker secretly relays and possibly alters the communication between two parties who believe they are directly communicating with each other. SOURCE: SP 800-63

manipulated variable attack: In a process that is intended to regulate some condition, a quantity or a condition that the control alters to initiate a change in the value of the regulated condition such as a setpoint.

manipulative communications deception: Alteration or simulation of friendly telecommunications for the purpose of deception. SOURCE: CNSSI-4009

manometer: An instrument that uses a column of liquid to measure pressure, although the term is currently often used to mean any pressure-measuring instrument. A vacuum gauge is used to measure the pressure in a vacuum—which is further divided into two subcategories: high and low vacuum (and sometimes ultra-high vacuum). SOURCE: Wikipedia

manual disconnector module: For manual disconnection of individual signals; useful during equipment start-up. Plugged between terminal socket and electronic module. For pluggable I/O modules only and not suitable for line voltage.

manual override switch: Manual override switches and potentiometers of output modules support direct operation. The positions of the manual override switches and potentiometers directly control outputs: independently. When a manual override switch or potentiometer is not in its default position ("auto"), the corresponding output LED will blink continuously, and the output module will send a feedback signal with the status "manual override" and the given override position to the Controller (which will then also store this information in its alarm memory).

manufacturing execution system (MES): A building control system that uses network computing to automate production control and process automation. By downloading recipes and work schedules and uploading production results, a MES bridges the gap between business and plant-floor or process-control systems.

manufacturing intelligence: Software used to bring a corporation's manufacturing-related data together from many sources for the purposes of reporting, analysis, visual summaries, and passing data between enterprise-level and plant-floor systems. As data is combined from multiple sources, it can be given a new structure or context that will help users find what they need regardless of where it came from. The primary goal is to turn large amounts of data into useful knowledge and drive business results.

Manufacturing Message Specification (MMS): A messaging protocol for transferring real time process data and supervisory control information between networked field devices and computer applications. (ISO 9506)

masquerade attack: A cyber-attack in which one system entity illegitimately poses as another entity. Also called a *spoofing attack*. SOURCE: SP 800-19

master-slave/token passing (MS/TP): Data link protocol as defined by the BACnet standard. Multiple speeds (data rates) are permitted by the BACnet MS/TP standard.

master terminal unit (MTU): See *SCADA server*.

mean swaps between failure (MSBF): A statistical calculation used to predict the average usefulness of a device without interruption of service.

mean time between failure (MTBF): A statistical calculation used to predict the average usefulness of a device without any interruption of service.

mean time to repair (MTTR): The average amount of time required to resolve most hardware or software problems with a given device.

meatspace: Or *meat world,* is the real world we live in, which contrast with the term *cyberspace*.

media transportation coverage: An insurance policy designed to cover transportation of items to and from a data center, the cost of reconstruction and the tracing of lost items. Coverage is usually extended to transportation and dishonesty or collusion by delivery employees.

medical device attacks: In 2012, a white hat hacker claimed he could kill a diabetic person from 300 feet away by ordering an insulin pump to deliver fatal doses of insulin. More recently, he announced he could hack pacemakers and implanted defibrillators.

media access control (MAC) address: A sublayer of the data link layer. Provides addressing and channel access control mechanisms that make it possible for several terminals or nodes to communicate within an Ethernet network. A link between the sublayer and the physical layer.

meets: When trains are running in opposite directions on a single-track railroad, meeting points (*"meets"*) are scheduled, at which each train must wait for the other at a passing place.

megabyte: Approximately one million bytes, 1,024 kilobytes.

memory scavenging: The collection of residual information from data storage. SOURCE: CNSSI-4009

menu map: Printed document supplied with controller illuminating menu item locations.

message digest: The result of applying a hash function to a message. Also known as a *hash value* or *hash output*. SOURCE: SP 800-107

message externals: Information outside of the message text, such as the header, trailer, and so forth. SOURCE: CNSSI-4009

mesh network: A network topology in which each node relays data for the network. All mesh nodes cooperate in the distribution of data in the network. Mesh networks can relay messages using either a *flooding* technique or a *routing* technique. With routing, the message is propagated along a path by *hopping* from node to node until it reaches its destination. The network is typically quite reliable, as there is often more than one path between a source and a destination in the network.

metamorphic and polymorphic malware attack: This category of malware keeps changing its code so each of its succeeding versions is different from the previous one. Metamorphic and polymorphic malware evades detection and conventional antivirus programs. It is difficult to write since it requires complicated techniques.

Metasploit Project: An open source SCADA security project that provides information about vulnerabilities to aid in penetration testing and development of NIDS signatures. Unlike other frameworks, Metasploit can also be used for anti-forensics. Expert programmers can write a piece of code exploiting a particular vulnerability, and test it with Metasploit to see if it gets detected. This process can be reversed technically—when a virus attacks using some unknown vulnerability, Metasploit can be used to test the patch for it.

microgrid: A localized network of electricity sources and loads that normally operates connected to the commercial power grid, but can function autonomously as needed. The smallest discrete network with the capability to operate independently is called a *nanogrid*. A nanogrid is usually a single building or a single energy domain.

mirroring: A method of storage in which data from one hard drive is duplicated on another hard drive so that both drives contain the same information, providing data redundancy.

misnamed files: A technique used to disguise a file's content by changing the file's name to something innocuous or altering its extension to a different type of file, forcing the examiner to identify the files by file signature versus file extension.

mission assurance category (MAC): The mission assurance category reflects the importance of information relative to the achievement of goals and objectives. Mission assurance categories are primarily used to determine the requirements for availability and integrity. The three defined mission assurance categories are:

- **MAC I**: Systems handling information that is determined to be vital to the operational readiness or mission effectiveness of deployed and contingency forces in terms of both content and timeliness. The consequences of loss of integrity or availability of a MAC 1 system are unacceptable and could include the immediate and sustained loss of mission effectiveness. MAC 1 systems require the most stringent protection measures.

- **MAC II**: Systems handling information that is important to the support of deployed and contingency forces. The consequences of loss of integrity are unacceptable. Loss of availability is difficult to deal with and can only be tolerated for a short time. The consequences could include delay or degradation in providing important support services or commodities that may seriously impact mission effectiveness or operational readiness.

- **MAC III**: Systems handling information that is necessary for the conduct of day-to-day business, but does not materially affect support to deployed or contingency forces in the short term. The consequences of loss of integrity or availability can be tolerated or overcome without significant impacts on mission effectiveness or operational readiness. The consequences could include the delay or degradation of services or commodities enabling routine activities.

mission critical: Any computer process that cannot be permitted to fail during normal business hours; some computer processes (e.g., telephone systems) must run all day long and require 100% uptime. Any telecommunications or information system that is defined as a national security system (Federal Information Security Management Act of 2002: FISMA) or processes any information the loss, misuse, disclosure, or unauthorized access to or modification of, would have a debilitating impact on the mission of an agency. SOURCE: SP 800-60

mobile code: Software programs or parts of programs obtained from remote systems, transmitted across a network, and executed on a local system without explicit installation or execution by the recipient. Some examples of software technologies that provide the mechanisms for the production and use of mobile code include Java, JavaScript, ActiveX, VBScript, and so forth. SOURCE: CNSSI-4009

mobile hot site: An 18-wheel truck preconfigured with backup computer equipment and peripherals delivered to the scene of cyber-attack. It is quickly connected to existing communication lines.

mobile ad hoc network (MANET): A continuously self-configuring, infrastructure-less network of mobile devices connected without wires. MANETs are a kind of Wireless ad hoc network that usually has a routable networking environment on top of a Link Layer ad hoc network. MANETs consist of a peer-to-peer, self-forming, self-healing network. MANETs typically communicate at radio frequencies (30 MHz: 5 GHz).

MANET cyber-attack: These cyber-attacks stop or slow the flow of information on the building automation network. MANETs are very vulnerable to malicious attacks.

mobile software agent: Programs that are goal-directed and capable of suspending their execution on one platform and moving to another platform where they resume execution. SOURCE: SP 800-19

Modbus: A serial protocol for control network communications used in utility control systems. It was originally published by Modicon (now Schneider Electric) for use with programmable logic controllers. It is a de facto standard and it is a common means of connecting industrial electronic devices. Data type names came about from its use in driving relays: a single-bit physical output is called a coil, and a single-bit physical input is called a *discrete input* or a *contact*. Modbus is a master/slave protocol, there is no way for a field device to "report by exception" (except over Ethernet TCP/IP, called *open-mbus*). Today Modbus is managed by the Modbus Organization.

Modbus Plus: A proprietary specification of Schneider Electric, normally implemented using a custom chipset. This is not a variant of Modbus. It is a different protocol, involving token passing.

modulator demodulator unit (MODEM): Is a network hardware device that modulates one or more carrier wave signals to encode digital information for transmission and demodulation that demodulates signals to decode the transmitted information. The goal is to produce a signal that can be transmitted easily and decoded to reproduce the original digital data. Modems can be used with any means of transmitting analog signals, from light emitting diodes to radio. A common type of modem is one that turns the digital data of a computer into modulated electrical signal for transmission over telephone lines and demodulated by another modem at the receiver side to recover the digital data. SOURCE: Wikipedia

monitoring: An ongoing activity that checks on the building control system, its users, or the environment.

monitoring and control (M&C) software: Front-end software that performs supervisory functions such as alarm handling, scheduling and data logging and provides a user interface for monitoring the building control system and configuring these functions. SOURCE: UFGS 25 10 10

mote: A sensor node in a wireless sensor network that is capable of performing some processing, gathering sensory information, and communicating with other connected nodes in the network. A mote is a node, but a node is not always a mote. Motes focus on providing the longest wireless range (dozens of km), the lowest energy consumption (a few uA) and the easiest development process for the user.

motion control: Motion control is a subfield of automation, in which the position or velocity of machines are controlled using some type of device such as a hydraulic pump, linear actuator, or electric motor, generally a servo. Widely used in the packaging, printing, textile, semiconductor production, and assembly industries.

motion control network: The network supporting the control applications that move parts in industrial settings, including sequencing, speed control, point-to-point control, and incremental motion.

mousetrapping: Software program that prevents a user from leaving a web site.

moving target defense: The presentation of a dynamic attack surface, increasing an adversary's work factor necessary to probe, attack or maintain presence in a cyber-target. SORCE: NICCS

monitoring and control software override report: Reports the points overridden by the monitoring and control software, including time overridden and identification of the operator (allegedly) overriding the point. SOURCE: UFGS 25 10 10

multifactor authentication: Authentication using two or more factors to achieve authentication. Factors include (1) something you *know* (e.g., password/PIN); (2) something you *have* (e.g., cryptographic identification device, token); or (3) something you *are* (e.g., biometric). SOURCE: SP 800-53

multi-homed: Network is directly connected to two ISPs.

multi-platform: The ability of a product or network to support a variety of computer platforms; also referred to as *cross-platform*.

CHAPTER 15

N

nanonetwork: A nanoscale network is a set of interconnected nanomachines (devices a few hundred nanometers or a few micrometers at most in size), which are able to perform only very simple tasks such as computing, data storing, sensing and actuation.

National Industrial Security Program (NISP): Program outlining criteria for cleared facilities to safeguard classified information while performing work on contracts, programs, bids, or research and development efforts. SOURCE: Executive Order 12829

National Information Infrastructure: Nationwide interconnection of communications networks, computers, databases, and consumer electronics that make vast amounts of information available to users. It includes both public and private networks, the Internet, the public switched network, and cable, wireless, and satellite communications. SOURCE: CNSSI-4009

National Security Emergency Preparedness Telecommunications Services: Telecommunications services that are used to maintain a state of readiness or to respond to and manage any event or crisis (local, national, or international) that causes or could cause injury or harm to the population, damage to or loss of property, or degrade or threaten the national security or emergency preparedness posture of the United States. SOURCE: SP 800-53; CNSSI-4009; 47 C.F.R., Part 64, App A

National Security Information: Information that has been determined pursuant to Executive Order 12958 as amended by Executive Order 13292, or any predecessor order, or by the Atomic Energy Act of 1954, as amended, to require protection against unauthorized disclosure and is marked to indicate its classified status. SOURCE: SP 800-53A; SP 800-60; FIPS 200

National Vulnerability Database (NVD): The US government repository of standards-based vulnerability management data. This data enables automation of vulnerability management, security measurement, and compliance (e.g., FISMA). SOURCE: http://nvd.nist.gov/

natural threats: Events caused by nature causing disruptions to a building's operation.

near field communication (NFC): A set of protocols that enable two electronic devices, one of which is usually a portable device such as a smartphone, to establish radio data communication with each other by bringing them closer than, typically, 4 inches from each other.

Nessus: A proprietary comprehensive vulnerability scanner developed by Tenable Network Security. It is free of charge for personal use in a non-enterprise environment. Nessus specializes in compliance checks, IP scans, sensitive data searches, and web site scanning and aids in finding the "weak spots."

net-centric architecture: A complex system of systems composed of subsystems and services that are part of a continuously evolving, complex community of people, devices, information, and services interconnected by a network that enhances information sharing and collaboration. Subsystems and services

© Luis Ayala 2016
L. Ayala, *Cybersecurity Lexicon*, DOI 10.1007/978-1-4842-2068-9_15

may or may not be developed or owned by the same entity, and, in general, will not be continually present during the full life cycle of the system of systems. Examples of this architecture include service-oriented architectures and cloud computing architectures. SOURCE: SP 800-37

NetFlow sensor: Reports statistics on network flow to classify a communication in terms of volume of data transfer, length of the transmission and number of packets. Monitors network packets and aggregates them in groups of the same source IP, destination IP, source port, and destination port. As packets in the same flow arrive, the number of bytes of new packets are recorded and the flow is considered closed if there is no transmission for a specified time. The number of packets and total duration of the flow is computed and the quantities are used to build models that show how data is moving.

Netsparker: A robust web application scanner that will identify vulnerabilities and suggest remedial action. Can also help exploit SQL injection and LFI (local file induction). It has a command-line and GUI interface, works on Microsoft Windows.

network: Information system(s) implemented with a collection of interconnected components. Such components may include routers, hubs, cabling, telecommunications controllers, key distribution centers, and technical control devices. SOURCE: SP 800-53; CNSSI-4009

Network Access Control (NAC): A feature provided by some firewalls that allows access based on a user's credentials and the results of health checks performed on the telework client device. SOURCE: SP 800-41

Network Address Translation (NAT): A routing technology used by many firewalls to hide internal system addresses from an external network through use of an addressing schema. SOURCE: SP 800-41

network architecture: The basic layout of a building control systems, such as terminals and the paths between them.

network-attached storage (NAS): A disk array storage system that is connected directly to a building control system rather than to the network server (i.e., host attached); functions as a server in a client/server relationship, has a processor, an operating system or micro-kernel, and processes file I/O protocols such as SMB and NFS.

network configuration tool: Software used to configure a building control network and set device configuration properties. This software creates and modifies the control network database (LNS Database).

network congestion: Occurs when a link or node is carrying so much data that its quality of service deteriorates. Typical effects include queueing delay, packet loss or the blocking of new connections. Network protocols that use aggressive retransmissions to compensate for packet loss tend to keep building control systems in a state of network congestion. Networks using these protocols can exhibit two stable states under the same level of load. The stable state with low throughput is known as *congestive collapse*.

network front-end: Device implementing protocols that allow attachment of a computer system to a network. SOURCE: CNSSI-4009

network interface jack: Contains no "intelligence" or "logic"; they are "dumb" devices used for wiring termination and providing a place to connect test equipment.

network intrusion detection system (NIDS): A device or software that monitors for malicious activity and rule violations and reports incidences.

network outage: An interruption in building control system availability as a result of a communication failure affecting computer terminals, processors, or workstations.

network resilience: The ability of a building control system network to (1) provide continuous operation (i.e., highly resistant to cyber-attack and able to operate in a degraded mode if damaged); (2) recover effectively if a cyber-attack does occur; and (3) scale to meet rapid or unpredictable demands.

Network Service Provider (NSP): A company that provides the national or international packet-switching networks that carry Internet traffic; also called a *backbone operator*.

network services: In cybersecurity work, a person that installs, configures, tests, operates, maintains, and manages networks and their firewalls, including hardware (e.g., hubs, bridges, switches, multiplexers, routers, cables, proxy servers, and protective building control systems) and software that permits the sharing and transmission of transmissions of information to support the security of building control systems. SOURCE: NICCS

network sniffing: A passive technique that monitors network communication, decodes protocols, and examines headers and payloads for information of interest. It is both a review technique and a target identification and analysis technique. SOURCE: SP 800-115

network tap: A hardware device that provides a way to access the data flowing across a building control system network. Similar to a phone tap or a vampire tap.

Network Mapper (Nmap): A security scanner used to discover network hosts and services on a building control system network, thus creating a "map" of the building control system network. Nmap sends specially crafted packets to the target host and then analyzes the responses. The following are typical uses of Nmap:

- Auditing the security of a device or firewall by identifying the network connections that can be made to or through it.

- Identifying open ports on a target host in preparation for auditing.

- Network inventory, network mapping, maintenance, and asset management.

- Auditing the security of a network by identifying new servers.

- Generating traffic to hosts on a network.

- Find and exploit vulnerabilities in a network.

network weaving: Penetration technique in which different communication networks are linked to access a building control system to avoid detection and trace-back. SOURCE: CNSSI-4009

newbie: Newb or noob is hacker slang term for a novice or newcomer, or inexperienced.

Niagara Framework: A set of hardware and software specifications for buildings and utility systems controls owned by Tridium Inc. and licensed to multiple vendors. The framework consists of front end (M&C) software, web based clients, field level control hardware, and engineering tools. It is not adopted by a recognized standards body and does not use an open licensing model. It is well-supported by multiple HVAC vendors and considered a de-facto open standard. SOURCE: UFGS 25 10 10

Night Dragon attack: This virus targeted global oil companies with the aim of finding project details and financial information about oil and gas field exploration and bids.

node (or network node): (1) Any device that is directly connected to the network, usually through Ethernet cable. Nodes include file servers and shared peripherals; the name used to designate a part of a network. This may be used to describe one of the links in the network, or a type of link in the network (for example, Host Node or Intercept Node). (2) A device that communicates using the CEA-709.1-C protocol and is connected to a CEA-709.1-C network. SOURCE: UFGS 25 10 10

node address: The logical address of a node on the network, consisting of a Domain number, Subnet number and Node number. Note that the "Node number" portion of the address is the number assigned to the device during installation and is unique within a subnet. This is not the factory-set unique Node ID (see Node ID).

node ID: A unique 48-bit identifier assigned at the factory to each intelligent electronic device.

noise rejection: Eliminating the hum or buzz when audio power amplifiers or similar devices create noise around the inputs that is not properly rejected.

No-Lone Zone (NLZ): Area, room, or space that, when staffed, must be occupied by two or more appropriately cleared individuals who remain within sight of each other. SOURCE: CNSSI-4009

non-local maintenance: Maintenance activities conducted by individuals communicating through a network; either an external network (e.g., the Internet) or an internal network. SOURCE: SP 800-53

nonce: A value used in security protocols that is never repeated with the same key. For example, nonces used as challenges in challenge-response authentication protocols generally must not be repeated until authentication keys are changed. Otherwise, there is a possibility of a replay attack. Using a nonce as a challenge is a different requirement than a random challenge, because a nonce is not necessarily unpredictable. SOURCE: SP 800-63

nonessential function: Building activities or information that could be interrupted or unavailable indefinitely without significantly jeopardizing critical building operations.

nonessential records: Records or documents, which, if lost or damaged, will not materially impair the building's ability to operate properly.

non-repudiation: Protection against an individual falsely denying having performed a particular action. Provides the capability to determine whether a given individual took a particular action such as creating information, sending a message, approving information, and receiving a message. SOURCE: SP 800-53; SP 800-18

NT (Microsoft Windows NT): An operating system developed by Microsoft for high-performance processors and networked systems.

null route: A network route (routing table entry) that goes nowhere. Matching packets are dropped (ignored) rather than forwarded, acting as a kind of very limited firewall. Also called a *blackhole route*. The act of using null routes is often called *blackhole filtering*.

null session: An anonymous (unauthenticated) connection to a freely accessible network share called IPC$ on Windows-based servers.

numerical control (NC): The automation of machine tools that are operated by precisely programmed commands encoded on a storage medium, as opposed to controlled manually by hand wheels or levers, or mechanically automated by cams alone. Most NC today is computer (or computerized) numerical control (CNC), in which computers play an integral part of the control.

CHAPTER 16

O

obfuscated spam: An e-mail that has been designed to fool anti-spam e-mail filter software. For example, replacing a letter in the word Viagra, so it is written as V!agra.

object: A passive entity that contains or receives information. SOURCE: SP 800-27 Passive information system-related entity (e.g., devices, files, records, tables, processes, programs, domains) containing or receiving information. Access to an object implies access to the information it contains. SOURCE: CNSSI-4009 Passive information system-related entity (e.g., devices, files, records, tables, processes, programs, domains) containing or receiving information. Access to an object (by a subject) implies access to the information it contains. SOURCE: SP 800-53

occupancy schedule: The occupancy schedule drawing indicates the typical building occupancy information such as days of the week and time of day the building is occupied.

occupancy sensor/vacancy sensor: A lighting control device that detects occupancy of a space by a person and turns the lights on or off automatically to save energy, using infrared, ultrasonic or microwave technology.

original equipment manufacturer (OEM): A company that manufactures a given piece of hardware (unlike a value-added reseller, which changes and repackages the hardware).

off-host processing: A backup mode of operation in which operations can continue throughout a building control system network despite loss of communication with the host server.

offline attack: An attack where the hacker obtains some data (typically by eavesdropping on an authentication protocol run, or by penetrating a building control system and stealing security files) that he is able to analyze in a building control system of his own choosing. SOURCE: SP 800-63

offline processing: A backup mode of operation in which processing can continue manually or in batch mode if the on-line building control system becomes unavailable.

offsite storage facility: A secure location where backup hardware, software, data files, documents, equipment, or supplies are stored.

online attack: An attack against an authentication protocol where the Attacker either assumes the role of a Claimant with a genuine Verifier or actively alters the authentication channel. The goal of the attack may be to gain authenticated access or learn authentication secrets. SOURCE: SP 800-63

online systems: An interactive building control system supporting users over a network of computer terminals.

one-time password attack: A one-time password is a code issued by a small electronic device every 30 or 60 seconds that is valid for only one login session or transaction. Online thieves have created real-time Trojan horse programs that can issue transactions to a bank while the account holder is online, turning the one-time password into a huge vulnerability. SOURCE: FFIEC

© Luis Ayala 2016
L. Ayala, *Cybersecurity Lexicon*, DOI 10.1007/978-1-4842-2068-9_16

one-way hash algorithm: Hash algorithms which map arbitrarily long inputs into a fixed-size output such that it is very difficult (computationally infeasible) to find two different hash inputs that produce the same output. Such algorithms are an essential part of the process of producing fixed-size digital signatures that can both authenticate the signer and provide for data integrity checking (detection of input modification after signature). SOURCE: SP 800-49; CNSSI-4009

open-loop controller: A controller that does not use feedback to determine if its output has achieved the desired goal of the input. Also called a *non-feedback controller*.

open systems network: A building control system network comprised of equipment that conforms to industry standards of interoperability between different operating systems (e.g., Unix, Windows NT).

operational controls: The security controls (i.e., safeguards or countermeasures) for an information system that primarily are implemented and executed by people (as opposed to systems). SOURCE: SP 800-53; SP 800-37; FIPS 200

operational exercise: Action-based exercise where staff rehearses reactions to a cyber-attack, drawing on their understanding of plans and procedures, roles, and responsibilities.

operational vulnerability information: Information that describes the presence of an information vulnerability within a specific operational setting or network. SOURCE: CNSSI-4009

operations technology: The hardware and software systems used to operate industrial control devices.

operating software: A type of building control system software supervising and directing all of the other software components plus the computer hardware.

operating system: The master control program (e.g., Windows) that manages a computer's internal functions and provides a means of control to the computer's operations and file system. An integrated collection of service routines for supervising the sequencing of programs by a computer. An operating system may perform the functions of input/output control, resource scheduling, and data management. It provides application programs with the fundamental commands for controlling the computer.

operating system (OS) fingerprinting: Analyzing characteristics of packets sent by a target, such as packet headers or listening ports, to identify the operating system in use on the target. SOURCE: SP 800-115

operations security (OPSEC): A process of identifying critical information and subsequently analyzing friendly actions attendant to military operations and other activities to: (a) identify those operations that can be observed by adversary intelligence systems; (b) determine what indicators adversary intelligence systems might obtain that could be interpreted or pieced together to derive critical information in time to be useful to adversaries; and (c) select and execute measures that eliminate or reduce to an acceptable level the vulnerabilities of friendly actions to adversary exploitation. SOURCE: JP 1-02, Department of Defense Dictionary of Military and Associated Terms

operator interface: The building control system command and information center. With it, data can be entered and displayed. Information such as current temperature values, control status, and so forth, can also be displayed. Typically, menu-driven, backlit LCD graphic display with clearly marked keys, makes the device easy to use. Touch-panel operation screens allow for easy operation by fingertip. User-configurable fast-access lists can contain selected data points, time programs, and parameters permitting customer-oriented operation.

organization chart: A diagram representative of the hierarchy of an organization's personnel.

organization-wide: A policy or function applicable to the entire organization.

outage: *See systems outage.*

outsider threat: An unauthorized entity outside the security domain that has the potential to harm an information system through destruction, disclosure, modification of data, and/or denial-of-service. SOURCE: CNSSI-4009

outsourcing: Transfer of data processing functions to an independent third party.

outstations: Remote computers or controllers for the electric utility industry and water companies.

overcurrent: In an electric power system, when a larger than intended electric current exists through a conductor, leading to excessive generation of heat, and the risk of fire or damage to equipment. Possible causes for overcurrent include short circuits, excessive load, incorrect design, or a ground fault. Fuses, circuit breakers, temperature sensors and current limiters are commonly used protection mechanisms to control the risks of overcurrent.

overload attack: In an overload cyber-attack, a shared resource or service is overloaded with requests to such a point that it's unable to satisfy requests from other users.

override: To change the value of a point outside of the normal sequence of operation where this change has priority over the sequence. An override can be accomplished in one of two ways: the point itself may be Commandable and written to with a priority or there may be a separate point on the controller for the express purpose of implementing the override.

overridable: A point is overridable (or commandable) if its Present_Value property can be changed using network variables outside of the normal sequence of operations where this change has priority over the sequence. SOURCE: UFGS 25 10 10

Overt Channel: Communications path within a computer system or network designed for the authorized transfer of data. SOURCE: CNSSI-4009

Overt Testing: Security testing performed with the knowledge and consent of the organization's IT staff. SOURCE: SP 800-115

Owner: The individual designated as being responsible for the protection of IT resources. The owner generally falls into one of two broad categories: custodial and owner. For example, the "owner" of the resources, may be the manager of the facility. Resources located within user areas may be "owned" by the manager of those areas. To assist with the determination of ownership, individual building control system boundaries must be established. A building control system is identified by logical boundaries being drawn around the various processing, communications, storage, and related resources. They must be under the same direct management control with essentially the same function, reside in the same environment, and have the same characteristics and security needs. Ownership of information and/or information processing resources may be assigned to an organization, subordinate functional element, a position, or a specific individual. When ownership is assigned to an organizational or functional element, the head of the unit so designated will be considered the resource owner. Some, but not necessarily all factors to be considered in the determination of ownership are:

- The originator or creator of the data.

- The organization or individual with the greatest functional interest.

- Physical possession of the resource.

CHAPTER 17

P

packet: A packet is the unit of data that is routed between an origin and a destination on the Internet or any other packet switched network.

packet analyzer: Software that observes and records network traffic. SOURCE: SP 800-61; CNSSI-4009

packet drop attack: A type of denial-of-service attack in which a router that is supposed to relay packets instead discards them. This usually occurs from a router becoming compromised from a number of different causes. One cause mentioned in research is through a denial-of-service attack on the router using a known DDoS tool. Because packets are routinely dropped from a lossy network, the packet drop attack is very hard to detect and prevent. Also called a *black hole attack*. The malicious router can also accomplish this attack selectively; for example, by dropping packets for a particular network destination, at a certain time of the day, a packet every n packets or every t seconds, or a randomly selected portion of the packets. This is called a *gray hole attack*.

packet sniffer: Software that observes and records network traffic. SOURCE: CNSSI-4009

page jacking: Stealing content such as source code from a web site and copying it to another web site.

pairing and bonding: Many services offered over Bluetooth can expose private data or let a connecting party control the Bluetooth device. Security reasons make it necessary to recognize specific devices, and thus enable control over which devices can connect to a given Bluetooth device. At the same time, it is useful for Bluetooth devices to be able to establish a connection without user intervention (for example, as soon as in range). To resolve this conflict, Bluetooth uses a process called *bonding*, and a bond is generated through a process called *pairing*. The pairing process is triggered either by a specific request from a user to generate a bond (for example, the user explicitly requests to "Add a Bluetooth device"), or it is triggered automatically when connecting to a service where (for the first time) the identity of a device is required for security purposes. These two cases are referred to as *dedicated bonding* and *general bonding*, respectively.

parallel test: A test of cyber-recovery procedures in which the objective is to parallel an actual building operations cycle.

parasitic Wi-Fi: It is possible to induce parasitic signals on the audio front end of voice-command-capable devices such as the iPhone. A hacker can send radio waves to any Android or iPhone that has Google Now or Siri enabled. The hack uses the phone's headphone cord as an antenna to convert electrical signals that appear to the phone's operating system to be audio coming from the microphone. Anything you can do through the voice interface you can do remotely and discretely through electromagnetic waves.

parity data: A block of information mathematically created from several blocks of user data to allow recovery of user data contained on a hard drive that has failed in RAID levels 3 and 5.

© Luis Ayala 2016
L. Ayala, *Cybersecurity Lexicon*, DOI 10.1007/978-1-4842-2068-9_17

Pass the Hash Attack: A hacking technique that allows an attacker to authenticate to a remote server/ service by using the underlying NTLM and/or LanMan hash of a user's password, instead of requiring the associated plaintext password as is normally the case. The attack exploits an implementation weakness in the authentication protocol in that the password hashes are not salted, and therefore remain static from session to session until the password is next changed.

passivation: The formation of a protective, passive, carbonate layer on galvanized steel surfaces to provide protection from corrosion.

passive attack: An attack against an authentication protocol where the attacker intercepts data traveling along the network between the claimant and verifier, but does not alter the data (i.e., eavesdropping). SOURCE: SP 800-63

passive Wi-Fi: A new type of Wi-Fi hardware that uses a transmission technique that's ultra-low powered (10,000 times less power than traditional Wi-Fi networking equipment). With Passive Wi-Fi, just one device produces a radio frequency. That frequency is relayed to your Wi-Fi-enabled device via separate, passive sensors that have only the baseband chip and an antenna and require almost no power.

passive wiretapping: The monitoring or recording of data while it is being transmitted over a communications link, without altering or affecting the data. SOURCE: CNSSI-4009

password: A string of characters (letters, numbers, and other symbols) used to authenticate an identity or to verify access authorization. SOURCE: FIPS 140-2

password security: A long password using symbols is stronger than just adding uppercase characters or numbers. Password attack software takes advantage of the fact that people tend to use uppercase characters at the start of passwords and numbers at the end.

patch: A fix for a software program where the actual binary executable and related files are modified. SOURCE: SP 800-123

patch management: The systematic notification, identification, deployment, installation, and verification of operating system and application software code revisions. These revisions are known as *patches*, *hot fixes*, *bug fixes* and *service packs*. SOURCE: CNSSI-4009

path histories: Maintaining an authenticatable record of the prior platforms visited by a mobile software agent, so that a newly visited platform can determine whether to process the agent and what resource constraints to apply. SOURCE: SP 800-19

payment card skimmer attack: Illegally installed device to read credit cards as customers pay.

pellistor: Registered trade name for a commercial device with a very small sensing element used in catalytic sensors and sometimes also called a *bead* or a *siegistor*.

penetration test (pen test): A test methodology in which assessors, using all available documentation (e.g., system design, source code, manuals) and working under specific constraints, attempt to circumvent the security features of an information system. Security testing in which evaluators mimic real-world attacks in an attempt to identify ways to circumvent the security features of an application, system, or network. Penetration testing often involves issuing real attacks on real systems and data, using the same tools and techniques used by actual attackers. Most penetration tests involve looking for combinations of vulnerabilities on a single system or multiple systems that can be used to gain more access than could be achieved through a single vulnerability. The tools used for pen-testing can be classified into two kinds—scanners and attackers. Some software/tools will show you the weak spots, some that show and attack. SOURCE: SP 800-53A, SP 800-115

periods processing: The processing of various levels of classified and unclassified information at distinctly different times. Under the concept of periods processing, the building control system must be purged of all information from one processing period before transitioning to the next. SOURCE: CNSSI-4009

permutation: Device used in cryptographic equipment to change the order in which the contents of a shift register are used in various nonlinear combining circuits. SOURCE: CNSSI-4009

personal computer interconnect (PCI): An industry-standard bus used in PCs, workstations, and servers.

personally identifiable information (PII): Information that can be used to distinguish or trace an individual's identity, such as their name, Social Security number, biometric records, and so forth, alone or combined with other personal or identifying information that is linked or linkable to a specific individual, such as date and place of birth, mother's maiden name, and so forth. SOURCE: CNSSI-4009

personal protective equipment (PPE): Protective clothing, helmets, goggles or other garments or equipment designed to protect the wearer's body from injury by blunt impacts, electrical hazards, heat, chemicals and infection.

personnel (security) clearance (PCL): An administrative determination that an individual is eligible, from a security point of view, for access to classified information of the same or lower category as the level of the personnel clearance being granted. SOURCE: NISPOM 2-104

peripheral equipment: Devices connected to a computer processor that perform such auxiliary functions as communications, data storage, printing, and so forth.

petabyte: 1,024 terabytes.

phantom power: A means of distributing a DC current through audio cables to provide power for microphones and other equipment. The supplied voltage is usually between 12 and 48 volts, with 48V being the most common.

pharming attack: A sophisticated MITM cyber- attack intended to redirect a web site's traffic to another, fake site. Pharming can be conducted either by changing the hosts file on a victim's computer or by exploitation of a vulnerability in DNS server software. DNS servers are computers responsible for resolving Internet names into their real IP addresses. Compromised DNS servers are sometimes referred to as *poisoned*. Pharming requires unprotected access to target a computer, such as altering a customer's home computer, rather than a corporate business server.

phishing attack: Tricking individuals into disclosing sensitive personal information by claiming to be a trustworthy entity in an electronic communication (e.g., Internet web sites). A digital form of social engineering that uses authentic-looking—but bogus—e-mails to request information from users or direct them to a fake web site that requests information. SOURCE: SP 800-83, SP 800-115

photo eye: A device used to detect the distance, absence, or presence of an object by using a light transmitter, often infrared, and a photoelectric receiver. They are used extensively in industrial manufacturing. There are three different functional types: opposed (through beam), retro-reflective, and proximity-sensing (diffused). Also called a *photo sensor*.

phreaking: A slang term describing the activity of a culture of people who study, experiment with, or explore telecommunication systems, such as equipment and systems connected to public telephone networks.

physical safeguards: Security safeguards or countermeasures to avoid, detect, counteract, or minimize security risks to physical property, information, building control systems, or other assets. Controls help to reduce the risk of damage or loss by stopping, deterring, or slowing down an attack against an asset.

physically isolated network: A network that is not connected to entities or systems outside a physically controlled space. SOURCE: SP 800-32

piconet: A small Bluetooth network created on an ad hoc basis that includes two or more devices. SOURCE: SP 800-121

pigging: The practice of using devices known as *pigs* to perform various maintenance operations on a pipeline. The devices are also known as *scrapers* or *go-devils*. Launched from pig-launcher stations, they travel through the pipeline to be received at any other station down-stream. *Smart pigs* (also known as *intelligent* or *intelligence pigs*) are used to detect anomalies in the pipe such as dents, metal loss caused by corrosion, cracking or other mechanical damage.

piggybacking attack: Intentional access of an open Wi-Fi network without harmful intent. Some jurisdictions prohibit it, some permit it, and others are not well defined. In the United States, the laws vary widely between states. For example, it is a third-degree felony in the state of Florida. New York law is the most permissive. The statute against unauthorized access only applies when the network "is equipped or programmed with any device or coding system, a function of which is to prevent the unauthorized use of said computer or computer system." In other words, the use of a network would only be considered unauthorized and illegal if the network owner had enabled encryption or password protection and the user bypassed this protection, or when the owner has explicitly given notice that use of the network is prohibited. A customer of a business providing hotspot service, such as a hotel or café, is generally not considered to be piggybacking.

pilot control module: An electronic device that controls the pilot light on a gas water heater, furnace or boiler.

Ping of Death Attack: A cyber-attack that sends a large echo request packet with to overflow the input buffers of the building control system causing it to crash.

ping scan attack: A passive cyber-attack looking for machines responding to pings.

ping sweep attack: A cyber-attack that pings a range of IP addresses, with the goal of finding building control system hosts that can be probed for vulnerabilities.

piping and instrumentation diagram (P&ID): A diagram in the process industry that shows the piping of the process flow together with the installed equipment and instrumentation.

PlaceRaider: Novel Trojan horse *visual malware* app that allows a hacker to engage in remote reconnaissance through use of a smartphone's camera and other sensors to obtain geolocation data and its accelerometer to create a 3D map of the phone's surroundings. A hacker can download images of the physical space, study the environment and carefully construct a three dimensional model of indoor environments to surveil the target's private home or work space. PlaceRaider can be used to steal virtual objects from the environment such as financial documents, information on computer monitors, and personally identifiable information.

plant supervision system: A plant supervision system allows managers to constantly monitor the performance and functioning of equipment and processes in an industrial plant giving personnel the ability to evaluate production output and dynamics, instantly identify bottlenecks from any location in real time to adjust operational parameters. A specific type of SCADA system.

platform: A computer hardware standard such as IBM, Sun, or Macintosh.

points: Building control systems have many data elements called *points*. Each point is a data monitor or sensor and these points can be hard or soft. A hard point can be an actual device. A soft point can be viewed as an application of software-calculated value used as a control output. Sometimes called a *tag*.

Point, Calculated: A value within the M&C Software that is not a network point but has been calculated by logic within the software based on the value of network points or other calculated points. Calculated points are sometimes called *virtual points* or *internal points*.

Point, Network: A value that the M&C Software reads from or writes to on a building control system network.

points schedule: The Points Schedule indicates Device address and NodeID, Input and Output SNVTs including SNVT Name, Type and Description, Hardware I/O, including Type (AI, AO, BI, BO) and Description.

Poison Ivy: A remote access tool (RAT) designed to scan for publicly known vulnerabilities in building control system and applications.

poison resistant: Capability of a catalytic sensor to reduce the effect of inhibiting substances or contaminants, such as silicones.

poison reverse attack: Split horizon with poisoned reverse includes routes in the updates, but sets the metrics to infinity (16) thereby advertising that their routes are not reachable.

policeware: Software designed to police citizens by monitoring discussion and interaction of its citizens. Computer and network surveillance of computer activity and data stored on a hard drive, or data being transferred over computer networks such as the Internet. The monitoring is often carried out covertly and may be completed by governments, corporations, criminal organizations, or individuals. It may or may not be legal and may or may not require authorization from a court or other independent government agency. Computer and network surveillance programs are widespread today and almost all Internet traffic can be monitored for illegal activity.

polling: A device requesting data from another device.

polyinstantiation: The ability of a database to maintain multiple records with the same key. Used to prevent inference attacks.

polymorphism: Process by which malicious software changes to avoid detection.

portable electronic device (PED): Any nonstationary electronic apparatus with singular or multiple capabilities of recording, storing, and/or transmitting data, voice, video, or photo images. This includes laptops, personal digital assistants, pocket personal computers, palmtops, MP3 players, cellular telephones, thumb drives, video cameras, and pagers. SOURCE: CNSSI-4009

portable shell: An environmentally protected structure that can be transported to a cyber-attack site so equipment can be available and installed quickly near the original location.

ports:

- Hardware: An equipment outlet for connecting cables.

- Network: An interface for communicating with a computer program over a network.

- Software: Software that has been changed to work on another operating system.

port scanning attack: Using a program to remotely determine which ports on a building control system are open (e.g., whether building control systems allow connections through those ports). SOURCE: SP 800-61; CNSSI-4009

PortScanDiggity: Passive port scanning via Google.

poser: A wannabe; a phrase used among crackers, phreaks and warez d00dz. Not as negative as *lamer* or *leech*. Not hacker slang.

potentially unwanted programs (PUP) attack: Installation of unwanted software in your building control system including search engines and toolbars. They are less harmful, but more annoying malware and include spyware, adware, as well as dialers.

power distribution unit (PDU): A PDU or *mains distribution unit* (MDU) is a device fitted with multiple outputs designed to distribute electric power, especially to racks of computers and networking equipment located within a building control system.

power outage: A short- or long-term loss of the electric power to an area (also called a *power cut, power blackout, power failure*, or *blackout*). Power outages are categorized relating to the duration and effect of the outage:

- A **permanent fault** is a massive loss of power typically caused by a fault on a power line. Power is automatically restored once the fault is cleared.

- A **brownout** is a drop in voltage in an electrical power supply. The term *brownout* comes from the dimming experienced by lighting when the voltage sags. Brownouts can cause poor performance of equipment or even incorrect operation.

- A **blackout** is the total loss of power to an area and is the most severe form of power outage that can occur. Blackouts that result from or result in power stations tripping are particularly difficult to recover from quickly. Outages may last from a few minutes to a few weeks depending on the nature of the blackout and the configuration of the electrical network.

Power over Ethernet (PoE) hack: Technology that uses unused conductors on Ethernet cabling to power low voltage devices. Up to 44 volts 350 ma is available. POE Plus can provide up to 25.5 Watts. An attacker that hacks into a security network and causes a power surge on the Ethernet cabling may be able to cause devices to fail.

power quality: While "power quality" is a convenient term, it is the quality of the voltage—rather than power or electric current—that is described. Power is simply the flow of energy and the current demanded by a load is largely uncontrollable. Electrical power can "swell" (exceed the nominal voltage by 10 to 80% for 0.5 cycle) or "sag" (voltage is below the nominal voltage by 10 to 90% for 0.5 cycle). Abrupt, very brief increases in voltage are called *spikes* and *undervoltage* occurs when the nominal voltage drops below 90% for more than 1 minute.

precursor: An observable occurrence or sign that an attacker may be preparing to cause an incident. SOURCE: SP 800-61

predisposing condition: A condition that exists within an organization, a mission/business process, enterprise architecture, or information system including its environment of operation, which contributes to (i.e., increases or decreases) the likelihood that one or more threat events, once initiated, will result in undesirable consequences or adverse impact to organizational operations and assets, individuals, other organizations, or the Nation. SOURCE: SP 800-30

preimage attack: In cryptography, a preimage attack on cryptographic hash functions tries to find a message that has a specific hash value. A cryptographic hash function should resist attacks on its preimage. Some significant preimage attacks have already been discovered, but they are not yet practical. If a practical preimage attack is discovered, it would drastically affect many Internet protocols. In this case, "practical" means that it could be executed by an attacker with a reasonable amount of resources (one that costs a few thousand dollars and takes a few weeks might be very practical).

pressure sensors: Wet/wet differential pressure sensors provide reliable, accurate measurement and control of proper applications, including the monitor and control of pump differential pressure, chiller/boiler differential pressure drop, and CW/HW system differential pressure. Used to measure pressure across pumps, filters, heat exchangers, compressors and other non-corrosive wet media applications. Duct static pressure sensors measure static pressure in the HVAC ductwork. SOURCE: Honeywell

print suppression: Eliminating the display of characters in order to preserve their secrecy. SOURCE: CNSSI-4009

primary and secondary pumps: Water distribution systems use a primary or main system pump and small secondary circulating pumps on branch runs. The primary pump runs all the time, and secondary pumps cycle on and off to create independent zones.

private key: The secret part of an asymmetric key pair that is typically used to digitally sign or decrypt data. SOURCE: SP 800-63

privilege escalation attack: Privilege escalation describes a cyber-attack where an attacker with some level of restricted access is able to, without authorization, elevate their privileges or access level. So, for example, a standard computer user may be able to fool the system into giving them access to restricted data; or even to "become root" and have full, unrestricted access to a system.

probe: (1) A device connected to the controller that monitors or measures a characteristic value in the fluid, like the conductivity. (2) A technique that attempts to access a system to learn something about the system. SOURCE: CNSSI-4009

probing attack: To attempt to connect to well-known services that may be running on a building control system; done to see if the building control system exists, and potentially to identify the software it is running.

procedural safeguards: Procedural measures taken to prevent a cyber-attack, such as frequent password changes, equipment inspections, random drills, security awareness programs, records retention programs, and so forth.

process control: Engineering discipline that deals with architectures, mechanisms and algorithms for maintaining the output of a specific process within a desired range.

process controller: A proprietary building control system, typically rack-mounted, that processes sensor input, executes control algorithms, and computes actuator outputs. A process controller may either use feedback or it may be open loop, and control may be continuous or cause a sequence of discrete events. Processes can be characterized as one or more of the following forms:

- **Discrete**: Manufacturing, motion and packaging applications. Robotic assembly can be characterized as discrete process control. Most discrete manufacturing involves the production of discrete pieces of product, such as metal stamping.

- **Batch**: Some applications require that specific quantities of raw materials be combined in specific ways for particular durations to produce an end result. Examples are the production of food, beverages and medicine.

- **Continuous**: Often, a physical building control system is represented through variables that are smooth and uninterrupted in time. The control of the water temperature in a heating jacket, for example, is an example of continuous process control.

- **Hybrid applications**: These have elements of discrete, batch and continuous process control.

process flow diagram (PFD): A diagram indicating the general flow of plant processes and equipment. The PFD displays the relationship between major equipment of a plant facility and not minor details.

process interlock: Something that prevents incorrect operation or possible damage to the process or equipment.

Profibus: A standard for field bus communication in automation technology.

profiling: Measuring the characteristics of expected activity so that changes to it can be more easily identified. SOURCE: SP 800-61; CNSSI-4009

program ID: A number stored in a device that identifies the intended device usage, node manufacturer, functionality of device and transceiver used.

program infector attack: Malware that attaches itself to existing program files.

programmable communicating thermostat (PCT): A thermostat that can receive information wirelessly.

programmable logic controller (PLC): A digital computer used for automation of typically industrial electromechanical processes, such as control of machinery on factory assembly lines, building control systems, or light fixtures. PLCs are used in many machines, in many industries. PLCs are designed for multiple arrangements of digital and analog inputs and outputs, extended temperature ranges, immunity to electrical noise, and resistance to vibration and impact. Programs to control machine operation are typically stored in battery-backed-up or non-volatile memory. A PLC is an example of a "hard" real-time system since output results must be produced in response to input conditions within a limited time, otherwise unintended operation will result.

programming fluids: Any caffeinacious stimulant; essential for those all-night hacking runs. Also called *wirewater*. Not alcoholic beverages.

progressive collapse: When a primary structural element building fails, resulting in the failure of adjoining structural elements, which in turn causes further structural failure. Since the resulting damage in a progressive collapse is disproportionate to the original cause, the term "disproportionate collapse" is frequently used in engineering to describe this collapse type. The nine-story Alfred P. Murrah Federal Building in Oklahoma City collapsed due to a truck bomb that was detonated outside the building. The bomb's compression wave caused floors 4 and 5 to shear up and off their columns and collapse on to floor 3. Floor 3 was connected to the main transfer beam, and pulled it inward when floors 4 and 5 fell on it. This caused all the vertical columns on the southern perimeter that were connected to the transfer beam to collapse as well, along with any floor sections that depended on those columns for vertical support. An analogous situation in the cyber realm would be where a relatively minor cyber-physical attack results in second-order effects that far exceed the severity of the initial attack.

promiscuous mode: A configuration setting for a network interface card that causes it to accept all incoming packets that it sees, regardless of their intended destinations. SOURCE: SP 800-94

property: A data element associated with an Object. Different Objects have different Properties, for example an Analog Input Object has a Present_Value Property (which provides the value of the underlying hardware analog input), a High_Limit Property (which contains a high limit for alarming), as well as other properties. SOURCE: UFGS 25 10 10

proprietary information: Material and information relating to or associated with a company's products, business, or activities, including but not limited to financial information; data or statements; trade secrets; product research and development; existing and future product designs and performance specifications; marketing plans or techniques; schematics; client lists; computer programs; processes; and know-how that has been clearly identified and properly marked by the company as proprietary information, trade secrets, or company confidential information. The information must have been developed by the company and not be available to the government or to the public without restriction from another source. SOURCE: CNSSI-4009

protected distribution system (PDS): Wire line or fiber optic system that includes adequate safeguards and/or countermeasures (e.g., acoustic, electric, electromagnetic, and physical) to permit its use for the transmission of unencrypted information through an area of lesser classification or control. SOURCE: CNSSI-4009, SP 800-53

protective technologies: Special tamper-evident features and materials employed for the purpose of detecting tampering and deterring attempts to compromise, modify, penetrate, extract, or substitute information processing equipment and keying material. SOURCE: CNSSI-4009

protocol: A standard that specifies the format of data and rules to be followed in data communication and network environments. A set of rules (i.e., formats and procedures) to implement and control some type of association (e.g., communication) between building control systems. SOURCE: CNSSI-4009

protocol analyzer: A device or software application that enables the user to analyze the performance of network data to ensure that the network and its associated hardware and software are operating within specifications.

protocol bridge: Translating one protocol to another, such as when TCP/IP traffic is converted to a proprietary control protocol such as Modbus, LonWorks, BACnet, and so forth.

protocol fuzzing attack: A testing technique used to generate valid and invalid packets with "random" header field values. The purpose is to analyze the behavior of a specific protocol by injecting unexpectedly malformed input parameter values. Random fuzzing is less effective, than *smart fuzzing* (tests based on the target specifications that require knowledge of the building control system).

Protocol Implementation Conformance Statement (PICS): A document, created by an equipment manufacturer that describes which portions of the BACnet standard are implemented by a given device.

proximity sensor: A non-contact sensor with the ability to detect the presence of an object within a specified range.

proxy: A proxy is an application that "breaks" the connection between client and server. The proxy accepts certain types of traffic entering or leaving a network and processes it and forwards it. This effectively closes the straight path between the internal and external networks making it more difficult for an attacker to obtain internal addresses and other details of the organization's internal network. Proxy servers are available for common Internet services; for example, a Hyper Text Transfer Protocol (HTTP) proxy used for web access, and a Simple Mail Transfer Protocol (SMTP) proxy used for e-mail. SOURCE: SP 800-44

proxy agent: A software application running on a firewall or on a dedicated proxy server that is capable of filtering a protocol and routing it between the interfaces of the device. SOURCE: CNSSI-4009

proxy server: A server that services the requests of its clients by forwarding those requests to other servers. SOURCE: CNSSI-4009

psig: Pressure in pounds per square inch gauge. Pipelines have a maximum rated operating pressure in psig, and many pipeline failures are associated with conditions that cause pressures to materially exceed that number.

push notification: A remote notifications feature. It is a highly efficient service for propagating information to intelligent, Internet-connected devices. Each device establishes an accredited and encrypted IP connection with the service and receives notifications over this persistent connection.

Pwnium: Google's bug bounty program that pays friendly hackers to find bugs in Google programs. It is run year-round, with a total prize pool of "∞ million." Google has paid as much as $2.7 million in one year.

CHAPTER 18

Q

quadrant: Short name referring to technology that provides tamper-resistant protection to cryptographic equipment. SOURCE: CNSSI-4009

qualitative assessment: Use of a set of methods, principles, or rules for assessing risk based on nonnumeric categories or levels. SOURCE: SP 800-30

quality of service: The measurable end-to-end performance properties of a network service, which can be guaranteed in advance by a Service-Level Agreement between a user and a service provider, so as to satisfy specific customer application requirements. These properties may include throughput (bandwidth), transit delay (latency), error rates, priority, security, packet loss, packet jitter, and so forth. SOURCE: CNSSI-4009

quantitative assessment: Use of a set of methods, principles, or rules for assessing risks based on the use of numbers where the meanings and proportionality of values are maintained inside and outside the context of the assessment. SOURCE: SP 800-30

quarantine: Place files containing malware in isolation for disinfection or future examination.

CHAPTER 19

R

rackmount: The cabinet that houses a server/storage workstation (also referred to as a *server rack*).

radiated noise: Noise from the radiation of another component, circuit, device, piece of equipment, building control system or object.

radio frequency identification (RFID): A form of automatic identification and data capture (AIDC) that uses electric or magnetic fields at radio frequencies to transmit information. SOURCE: SP 800-98

radiation monitoring: The process of receiving images, data, or audio from an unprotected source by searching for radiation signals.

Radio over IP (RoIP): The same concept as the Internet-based telephone system known as Voice over IP (VoIP). A RoIP connects to a router over a wireless network, it picks up your voice and sends it over Internet to other radios connected wirelessly to the Internet. Like VoIP, RoIP has unlimited range.

RAID Advisory Board (RAB): An organization of storage system manufacturers and integrators dedicated to advancing the use and awareness of RAID and associated storage technologies; started in 1992, RAB states its main goals as education, standardization and certification.

rainbow table: A precomputed table for reversing cryptographic hash functions, usually for cracking password hashes. A building control system that requires password authentication must contain a database of passwords, either hashed or in plaintext, and various methods of password storage exist. Because the tables are vulnerable to theft, storing the plaintext password is dangerous. Most databases store a cryptographic hash of a user's password in the database. No one: including the authentication system: can determine what a user's password is simply by looking at the value stored in the database. Instead, when a user enters his or her password for authentication, it is hashed and that output is compared to the stored entry for that user (which was hashed before being stored). If the two hashes match, access is granted.

ramp generator: A function generator that increases its output voltage up to a specific value, called a *ramp*. Among other uses, it is used in electrical generators or electric motors to avoid jolts when changing a load.

randomizer: Analog or digital source of unpredictable, unbiased, and usually independent bits. Randomizers can be used for several different functions, including key generation or to provide a starting state for a key generator. SOURCE: CNSSI-4009

ransomware: A type of malware that restricts access to a computer system that it infects in some way, and demands that the user pay a ransom to the operators of the malware to remove the restriction. Some forms of ransomware systematically encrypt files on the system's hard drive (*cryptoviral extortion*) using a large key that may be technologically infeasible to breach without paying the ransom, while some may simply lock the system and display messages intended to coax the user into paying. Ransomware typically propagates as a Trojan horse, whose payload is disguised as a seemingly legitimate file.

© Luis Ayala 2016
L. Ayala, *Cybersecurity Lexicon*, DOI 10.1007/978-1-4842-2068-9_19

Rapid Recovery Repairs (R³): Standby contract whereby a construction company is on call in the event of a cyber-physical attack on a critical facility, such as a hospital. The construction company is able to respond quickly because a contract is already in place to provide emergency utilities and commence repairs as soon as possible.

rapid rise error: Error message indicating boiler flow or return temperature water rising too rapidly.

real time: Immediate processing of input or notification of status.

real-time operating system (RTOS): An operating system (OS) intended to serve real-time application process data as it comes in, typically without buffering delays. Processing time requirements is measured in tenths of seconds or shorter. A key characteristic of an RTOS is the level of its consistency concerning the amount of time it takes to accept and complete an application's task; the variability is *jitter*. A *hard* real-time operating system has less jitter than a *soft* real-time operating system. The chief design goal is not high throughput, but rather a guarantee of a soft or hard performance category. An RTOS that can *usually* or *generally* meet a deadline is a soft real-time OS, but if it can meet a deadline deterministically it is a hard real-time OS.

real-time reaction: Immediate response to a penetration attempt that is detected and diagnosed in time to prevent access. SOURCE: CNSSI-4009

recovery: The activities after a cyber-attack to restore essential building services and operations in the short and medium term and fully restore all capabilities in the longer term.

reciprocal agreement: An agreement between two organizations with compatible computer configurations allowing either organization to utilize the other's excess processing capacity in the event of a cyber-attack.

record retention: Storing historical documentation for a set period of time, usually mandated by state and federal law or the Internal Revenue Service.

recovery action plan: The comprehensive set of documented tasks to be carried out during recovery operations.

recovery alternative: The method selected to recover the critical building functions following a cyber-attack.

recovery capability: This defines all of the components necessary to perform recovery after a cyber-attack. These components can include a plan, an alternate site, change control process, and controls network rerouting.

recovery management team: A staff responsible for development and maintenance of a cyber-physical attack recovery plan. Also responsible for declaring a cyber-attack and providing direction during the recovery process.

recovery planning team: A staff appointed to oversee the development and implementation of a cyber-physical attack recovery plan.

recovery point objective (RPO): The point in time to which data must be restored in order to resume processing transactions. RPO is the basis on which a data projection strategy is developed. SOURCE: SP 800-34

recovery procedures: The actions necessary to restore a building control system's processing capability and data files after a building control system failure. SOURCE: CNSSI-4009

recovery time: The overall length of time an information system's components can be in the recovery phase before negatively impacting the organization's mission or mission/business functions. SOURCE: SP 800-34

red signal: Any electronic emission (e.g., plain text, key, key stream, subkey stream, initial fill, or control signal) that would divulge national security information if recovered. SOURCE: CNSSI-4009

red/black concept: The careful segregation in cryptographic systems of signals that contain sensitive or classified plaintext information (red signals) from those that carry encrypted information, or ciphertext (black signals). Sometimes called the *red-black architecture* or *red/black engineering*. Encryption devices are often called *blackers*, because they convert red signals to black. Separation of electrical and electronic circuits, components, equipment, and systems that handle unencrypted information (red), in electrical form, from those that handle encrypted information (black) in the same form. SOURCE: CNSSI-4009

Red Team: A group of people authorized and organized to emulate a potential adversary's attack or exploitation capabilities against an enterprise's security posture. The Red Team's objective is to improve enterprise information assurance by demonstrating the impacts of successful attacks and by demonstrating what works for the defenders (i.e., the Blue Team) in an operational environment. SOURCE: CNSSI-4009

Red Team exercise: A cyber-attack exercise, reflecting real-world conditions, that is conducted as a simulated attempt by an adversary to attack or exploit vulnerabilities in an enterprise's building control systems. SOURCE: SP 800-53

redundancy: Additional or alternative building control systems, subsystems, assets, or processes that maintain a degree of overall functionality in case of loss or failure of a building control system, subsystem, asset, or process.

Redundant Array of Independent (or inexpensive) Disks (RAID): A collection of storage disks with a controller (or controllers) to manage the storage of data on the disks.

redundant control server: A backup to the building control system server that maintains the current state of the control server at all times.

redundant data path (RDP): Dot Hill's software technology that creates an alternate data path between the server and the storage system in the event of building control system component failures to ensure continuous access to data.

reference monitor: The security engineering term for IT functionality that

- controls all access
- cannot be bypassed
- is tamper-resistant
- provides confidence that the other three items are true

SOURCE: SP 800-33

regression analysis: Scripted tests to test software for all possible input it should expect.

reheat coil: In contrast to preheat elements that are located before the air handler's cooling coil, heating elements that are located downstream of an air handling system's cooling coils are referred to as being in the *reheat position*. The HVAC system works by having the AHU cool all the air just to the point needed by the hottest zone with its VAV box wide open for maximum air flow to the zone. The zones with lesser cooling demand will throttle their VAV boxes down as far as their minimum flows, and if the air is still too cold, will then reheat the air using a hot water or electric resistance coil at each zone. Reheat systems can be very energy intensive because they use simultaneous heating and cooling to achieve temperature (and sometimes humidity) control.

reintegration: The careful, methodical reconnection of devices on a segregated network to fully functioning operation.

relay: An electromechanical device that completes or interrupts an electrical circuit by physically moving conductive contacts. The resultant motion can be coupled to another mechanism such as a valve or a circuit breaker.

relocatable shell: *See portable shell*

remediation: The act of correcting a vulnerability or eliminating a threat. Three possible types of remediation are installing a patch, adjusting configuration settings, or uninstalling a software application. SOURCE: SP 800-40

remediation plan: A plan to perform the remediation of one or more threats or vulnerabilities facing an organization's building control systems. The plan typically includes options to remove threats and vulnerabilities and priorities for performing the remediation. SOURCE: SP 800-40

remnant files: Files left by hackers on compromised building control systems. These can be sniffer log files, password files, exploit scripts, and source code to various programs (may also be called *artifacts*).

remote access: Access to an organizational information system by a user (or an information system acting on behalf of a user) communicating through an external network (e.g., the Internet). SOURCE: SP 800-53 Access by users (or information systems) communicating external to an information system security perimeter. SOURCE: SP 800-18 The ability for an organization's users to access its nonpublic computing resources from external locations other than the organization's facilities. SOURCE: SP 800-46 Access to an organization's nonpublic information system by an authorized user (or an information system) communicating through an external, non-organization-controlled network (e.g., the Internet). SOURCE: CNSSI-4009

remote access tool (RAT): A piece of software that allows a remote "operator" to control a building control system as if he has physical access to that building control system. While desktop sharing and remote administration have many legal uses, RAT software is usually associated with criminal or malicious activity. Malicious RAT software is typically installed without the victim's knowledge, often as payload of a Trojan horse, and tries to hide its operation from the victim and from security software.

Such tools provide an operator the following capabilities:

- Screen/camera capture, image control or microphone control

- File management (download/upload/execute, etc.)

- Shell control (from command prompt)

- Computer control (power off/on/log off if remote feature is supported)

- Registry management (query/add/delete/modify)

- Hardware Destroyer (overclocker)

remote code execution vulnerability: Could enable an attacker to execute PHP code on a web server and bypass security mechanisms. Can allow the attacker to gain administrative access to the building control system.

remote diagnostics: Diagnostics activities conducted by individuals communicating external to a building control system security perimeter. This is **a very bad idea**. SOURCE: CNSSI-4009

remote I/O: A local area network designed to connect controllers to a variety of intelligent devices such as operator interfaces and AC or DC drives.

remote maintenance: Maintenance activities conducted by individuals communicating external to a building control system security perimeter. Also, *a very bad idea*. SOURCE: CNSSI-4009

remote terminal unit (RTU): A microprocessor-controlled electronic device that monitors analog and digital parameters and transmits data to the BAS Central Monitoring Station. A RTU monitors and transmits values as input or output signals from I/O devices such as meters, pressure transducers, pump starter auxiliary contacts, and so forth, from within the SCADA System. Signals created from a device such as a water meter and sent to the RTU are called *input signals*. Signals created within the RTU and sent elsewhere are called *output signals*. Signals are of the following types:

- **Digital**: ON/OFF discrete signal such as an equipment contact closure wired to the isolated inputs of the RTU and is generally read as 0 or 1 in value. These values could be a RUN status from a pump starter auxiliary contact, pressure switch, and so forth.

- **Analog**: A continuous signal that changes smoothly over a given range is brought into the RTU via a 4 to 20 milliamp signal. These are real values such as water levels, pressure or turbidity and are not discrete signals such as ON/OFF.

- **Counter**: Pulse signals from flow meter or similar occurrence meters that count the number of times an event occurs.

remote-to-local user (R2L) cyber-attack: When a hacker has the ability to send packets over a building control system network (but who does not have a valid user account) exploits a system vulnerability to gain access as a user.

repeater: Hardware device that connects two network segments and retransmits information received on one side to the other.

replay attack: Cyber-attack that involves capturing traffic sent over the network, and then reinjecting it again later, causing commands to be executed twice. A variety of mechanisms are designed to prevent replay attacks such as by using timestamps or session tokens. SOURCE: CNSSI-4009

repudiation attack: When a user denies that he or she performed an action or initiated a transaction. You need defense mechanisms in place to ensure that all user activity can be tracked and recorded. Otherwise, a user can simply deny having knowledge of the transaction or communication and later claim that such transaction or communication never took place.

reset: A common control function that attempts to return a building control system to its normal operating condition, canceling all active programming until reset is complete and new input status information can be processed. All input devices must be in their normal state before the building control system will reset completely.

residual risk: The remaining potential risk after all IT security measures are applied. There is a residual risk associated with each threat. SOURCE: SP 800-33 Portion of risk remaining after security measures have been applied. SOURCE: CNSSI-4009; SP 800-30

resilience: The ability to adapt to changing conditions and prepare for, withstand, and rapidly recover from disruption. SOURCE: SP 800-34

resource encapsulation: Method by which the reference monitor mediates accesses to an information system resource. Resource is protected and not directly accessible by a subject. Satisfies requirement for accurate auditing of resource usage. SOURCE: CNSSI-4009

resource exhaustion attack: Resource exhaustion cyber-attacks involve tying up limited resources on a building control system, making them unavailable to other users.

resource starvation: A condition where a computer process cannot be supported by available computer resources. Resource starvation can occur due to the lack of computer resources or the existence of multiple processes that are competing for the same computer resources.

response: The activities that address the short-term, direct effects of a cyber-physical attack and may also support short-term return of equipment to operational condition.

retrofit: The addition of new technology or features to older building control systems.

resolver: A rotary electrical transformer used for measuring degrees of rotation.

reversing valve: A type of valve in a heat pump that changes the direction of refrigerant flow so the heat pump refrigeration cycle is changed from cooling to heating or vice versa. This allows a single piece of equipment to heat or cool the facility with the same hardware.

riser diagram of building control network: The riser diagram indicates all DDC hardware and all network hardware, including network terminators (may be in tabular form). For each item, it provides a unique identifier, common descriptive name, physical sequential order (previous and next device on the network), room identifier and location within a room.

risk: The level of impact on organizational operations (including mission, functions, image, or reputation), organizational assets, or individuals resulting from the operation of an information system given the potential impact of a threat and the likelihood of that threat occurring. SOURCE: FIPS 200

risk analysis: The process of identifying the risks to system security and determining the likelihood of occurrence, the resulting impact, and the additional safeguards that mitigate this impact. Part of risk management and synonymous with *risk assessment*. SOURCE: SP 800-27

risk assessment: The process of identifying risks to agency operations (including mission, functions, image, or reputation), agency assets, or individuals by determining the probability of occurrence, the resulting impact, and additional security controls that would mitigate this impact. Part of risk management, synonymous with risk analysis. Incorporates threat and vulnerability analyses. SOURCE: SP 800-53; SP 800-53A; SP 800-37

risk management: The process of the identification, measurement, control, and minimization of security risk in building control system. Also, it means to assess risk, take actions to reduce risk to an acceptable level, and maintain risk at that level. Inherent in this definition are the concepts that risk cannot be completely eliminated and the most secure building control system is the one that is not turned on. SOURCE: SP 800-53; SP 800-53A; SP 800-37

robustness: The ability of a building or structure to withstand events like fire, explosions, impact or the consequences of human error, without being damaged to an extent disproportionate to the original cause. A structure designed and constructed to be robust should not suffer from progressive collapse under accidental loading. In situ cast concrete structures, are less susceptible to collapse than large-panel systems and precast concrete buildings. The ability of an information assurance entity to operate correctly and reliably across a wide range of operational conditions, and to fail gracefully outside of that operational range. SOURCE: CNSSI-4009

Robust Security Network (RSN): A wireless security network that only allows the creation of Robust Security Network Associations (RSNAs). SOURCE: SP 800-48

rogue device: An unauthorized node on a network. SOURCE: SP 800-115

rogue access point: A rogue access point is a wireless access point that has been installed on a secure network without explicit authorization from a local network administrator, whether added by a well-meaning employee or by a malicious insider. Although it is technically easy for a well-meaning employee to install a "soft access point" or an inexpensive wireless router: perhaps to make access from mobile devices easier: it is likely that they will configure this as "open," or with poor security, and potentially allow access to unauthorized parties. If an attacker installs a rogue access point they are able to run various types of vulnerability scanners, and rather than having to be physically inside the building, a hacker can attack remotely: perhaps from a reception area, adjacent building, or car parking lot.

rogue scanner: A network security tool to automatically discover rogue access points

rogue security software: A form of malicious software and Internet fraud that misleads users into believing there is a virus on their computer, and manipulates them into paying money for a fake malware removal tool (that actually introduces malware to the computer). It is a form of scareware that manipulates users though fear. It is also a form of ransomware.

role-based access control: Assigns control to users based on their functions and determines access based on those roles. SOURCE: SP 800-95

Ronan Point: A 22-story tower in East London, which partly collapsed on May 16, 1968, when a gas explosion demolished a load-bearing wall, causing the collapse of one entire corner of the building.

roof and tunnel hacking: The unauthorized physical exploration of roof and utility tunnel spaces such as by college students in campus buildings.

rootkit: A set of tools used by an attacker after gaining root-level access to a host to conceal the attacker's activities on the host and permit the attacker to maintain root-level access to the host through covert means. SOURCE: SP 800-61; CNSSI-4009

route flapping: A router that transmits routing updates alternately advertising a destination network first via one route, then via a different route.

router: An electronic device connecting two or more networks that routes incoming data packets to the appropriate network by retransmitting signals received from one subnet onto the other.

RPC scan: Determines which RPC services are running on a machine.

RS-232: A standard for serial communication transmission of data. Many intelligent devices have an RS-232 port built into the device for troubleshooting by maintenance personnel, or to install software upgrades or patches. USB has largely displaced RS-232 from most of its peripheral interface roles.

Rule Set Based Access Control (RSBAC): Targets actions based on rules for entities operating on objects.

rupture disc: A non-reclosing pressure relief device that, in most uses, protects a pressure vessel, equipment, or system from over pressurization or potentially damaging vacuum conditions, also known as a *pressure safety disc, burst disc, bursting disc,* or *burst diaphragm*.

CHAPTER 20

S

S-box: Nonlinear substitution table used in several byte substitution transformations and in the Key Expansion routine to perform a one-for-one substitution of a byte value. SOURCE: FIPS 197

S/MIME: A set of specifications for securing electronic mail. Secure/ Multipurpose Internet Mail Extensions (S/MIME) is based upon the widely used MIME standard and describes a protocol for adding cryptographic security services through MIME encapsulation of digitally signed and encrypted objects. The basic security services offered by S/MIME are authentication, non-repudiation of origin, message integrity, and message privacy. Optional security services include signed receipts, security labels, secure mailing lists, and an extended method of identifying the signer's certificate(s). SOURCE: SP 800-49

safe area: Work area in which there is no danger of contamination with explosive gases.

safeguards: The protective measures and controls that are prescribed to meet the security requirements specified for a building control system. Safeguards may include security features, management constraints, personnel security, and security of physical structures, areas, and devices. Synonymous with *security controls* and *countermeasures*. SOURCE: SP 800-53; SP 800-37; FIPS 200; CNSSI-4009

safety instrumented system: A building control system that uses control devices and sensors designed to put the system in a safe state when predetermined conditions are violated. Also called an *emergency shutdown system* (ESS), *safety shutdown system* (SSD), and *safety interlock system* (SIS).

safety interlock: Something that prevents possible damage or death to people operating that process or equipment.

safety relay: A convenient and economical solution for incorporating control reliability into a safety circuit.

salt: A non-secret value that is used in a cryptographic process, usually to ensure that the results of computations for one instance cannot be reused by an attacker. SOURCE: SP 800-63; CNSSI-4009

salvage and restoration: The process of reclaiming or refurbishing computer hardware, vital records, office facilities, and so forth, following a cyber-attack.

salvage procedures: Specified procedures to be activated if equipment or a facility should suffer a cyber-attack.

Samhain sensor: Frequently checks the critical building control system files for additions, modifications and deletions. All changes are immediately logged locally or reported to a remote log server. These include timestamps of changes, file names, violation type, and changes in the building control system kernel.

sample plan: A generic cyber-physical attack recovery plan that can be tailored to fit a particular organization.

© Luis Ayala 2016
L. Ayala, *Cybersecurity Lexicon*, DOI 10.1007/978-1-4842-2068-9_20

sandboxing: A method of isolating application modules into distinct fault domains enforced by software. The technique allows untrusted programs written in an unsafe language, such as C, to be executed safely within the single virtual address space of an application. Untrusted machine interpretable code modules are transformed so that all memory accesses are confined to code and data segments within their fault domain. Access to building control system resources can also be controlled through a unique identifier associated with each domain. SOURCE: SP 800-19

sanitization: Process to remove information from media such that information recovery is not possible. It includes removing all labels, markings, and activity logs. SOURCE: FIPS 200 A general term referring to the actions taken to render data written on media unrecoverable by both ordinary and, for some forms of sanitization, extraordinary means. SOURCE: SP 800-53; CNSSI-4009

satellite communication: Data communications via satellite. For geographically dispersed organizations, this may be a viable alternative to ground-based communications in the event of a cyber-physical attack.

scalable: The ability of a product or network to accommodate growth.

scams: Internet scams such as when scamsters phone computer users randomly and offer to fix non-existent computer problems on their computer for a fee.

scanning: Sending packets or requests to another system to gain information to be used in a subsequent attack. SOURCE: SP 800-61; CNSSI-4009

scanning attack: Any of the following:

- **active port scanning**: Actively send network packets to enumerate all open ports of a device, including both TCP and UDP.

- **passive traffic mapping/scanning**: Passively record network traffic. Discover ports that are normally used, without detecting open ports not actively used.

- **version scanning**: Actively attempt to discover the protocol by connecting to open ports.

- **vulnerability scanning**: Actively connect to a remote device and exploit known vulnerabilities.

SCADA (supervisory control and data acquisition): A system that operates with coded signals over communication channels to provide control of remote equipment (using typically one communication channel per remote station). The control system may be combined with a data acquisition system by adding the use of coded signals over communication channels to acquire information about the status of the remote equipment for display or for recording functions. It is a type of *industrial control system* (ICS). Industrial control systems are computer-based systems that monitor and control industrial processes that exist in the physical world. SCADA systems historically distinguish themselves from other ICS systems by being large-scale processes that can include multiple sites, and large distances. These processes include industrial, infrastructure, and facility-based processes, described as follows:

- Industrial processes include those of manufacturing, production, power generation, fabrication, and refining, and may run in continuous, batch, repetitive, or discrete modes.

- Infrastructure processes may be public or private, and include water treatment and distribution, wastewater collection and treatment, oil and gas pipelines, electrical power transmission and distribution, wind farms, civil defense siren systems, and large communication systems.

- Facility processes occur both in public facilities and private ones, including buildings, airports, ships, and space stations. They monitor and control heating, ventilation, and air conditioning systems (HVAC), access, and energy consumption.

SCADA duration surface: Unlike most IT equipment found in a corporate network that is normally replaced every 2 to 3 years, a SCADA system typically has a "duration surface" of 25 years. This makes SCADA systems more vulnerable to persistent threats, allowing more time to develop exploits against these slow-changing systems.

SCADA server: The device that acts as the master controller in a SCADA system.

scareware: A form of malicious software that uses social engineering to cause shock, anxiety, or the perception of a threat in order to manipulate users into buying unwanted software. Scareware is part of a class of malicious software that includes rogue security software, ransomware and other scam software with malicious payloads, which have limited or no benefit to users, and are pushed by unethical marketing practices. Some forms of spyware and adware also use scareware tactics. A tactic frequently used by criminals involves convincing users that a virus has infected their computer, then suggesting that they download (and pay for) fake antivirus software to remove it. Usually the virus is entirely fictional and the software is non-functional or malware itself.

scatternet: A type of ad hoc computer network consisting of two or more piconets. A chain of piconets created by allowing one or more Bluetooth devices to each be a slave in one piconet and act as the master for another piconet simultaneously. A scatternet allows several devices to be networked over an extended distance. SOURCE: SP 800-121

scavenging attack: Unauthorized searching through data in a BCS, ICS, or SCADA system to gain knowledge of sensitive data. Searching through object residue to acquire data. SOURCE: CNSSI-4009

scope: Predefined areas of operation for which a cyber-attack recovery plan is developed.

screen scraper: A virus or physical device that logs information sent to a computer display to capture information.

scripts: A piece of code that is loaded and run by your browser. The most common types are Java, HTML, and Flash-based plug-ins.

script kiddie: In hacking culture, a script kiddie is an unskilled individual who uses scripts or programs developed by others to attack networks. Also known as a *skiddie, skid, script bunny, script kitty*, and *skidiot*.

SearchDiggity: The primary cyber tool of the Google Hacking Diggity Project.

secret key: A cryptographic key that is used for both encryption and decryption, enabling the operation of a symmetric key cryptography scheme. SOURCE: FIPS 140-2

sectionalizer: A downstream protective device for electric power distribution with a tripping mechanism triggered by a counter or a timer. It detects and counts fault current interruptions by the recloser (or circuit breaker). After a predetermined number of interruptions, the sectionalizer will open, thereby isolating the faulty section of the circuit, allowing the recloser to restore supply to the other non-fault sections.

secure: In computer terminology, a secure LAN or a secure device means that the routing addresses on the network are monitored and allowed to proceed only for authorized users. This network traffic monitoring and authorization process is referred to as owner's *firewalls*. Building control systems and devices, when not being monitored, are referred to as being outside of owner's secure firewall and the term *non-secure* is applied.

secure erase: An overwrite technology using firmware-based process to overwrite a hard drive. A drive command defined in the ANSI ATA and SCSI disk drive interface specifications, which runs inside drive hardware. It completes in about 1/8 the time of 5220 block erasure. SOURCE: SP 800-88

secure configuration: Restricting the functionality of every device, operating system, and application to the minimum needed for the building control system to operate properly. A secure configuration minimizes the information that Internet-connected devices disclose about their configuration and software version and ensure they cannot be probed for vulnerabilities.

secure state: Condition in which no subject can access any object in an unauthorized manner. SOURCE: CNSSI-4009

security: A condition that results from the establishment and maintenance of protective measures that enable an enterprise to perform its mission or critical functions despite risks posed by threats to its use of information systems. Protective measures may involve a combination of deterrence, avoidance, prevention, detection, recovery, and correction that should form part of the enterprise's risk management approach. SOURCE: CNSSI-4009

security appliance: A server that is designed to protect building control system networks from unwanted traffic. It is a simple and cost-effective way to segment a building control system network into security zones. The user defines rules that specify exactly which devices are allowed to communicate, what protocols they may use, and what actions those protocols perform. Any communication that is not on the "allowed" list is automatically blocked and reported. The Tofino Security Appliance is an excellent example and *a very good idea.*

security attribute: A security-related quality of an object. Security attributes may be represented as hierarchical levels, bits in a bit map, or numbers. Compartments, caveats, and release markings are examples of security attributes. SOURCE: FIPS 188

security audit: Independent review and examination of a BCS's records and activities to determine the adequacy of a building control system's controls, ensure compliance with established security policy and procedures, detect breaches in security services, and recommend changes that are needed for countermeasures.

security automation: The use of information technology in place of manual processes for cyber-physical attack response and management.

security banner: A banner at the top or bottom of a computer screen that states the overall classification of the system in large, bold type. Also can refer to the opening screen that informs users of the security implications of accessing a computer resource. SOURCE: CNSSI-4009

security by obscurity: False sense of security when managers believe digital control systems like SCADA are not too complex to be hacked by foreigners attempting to decipher logic strings or icons on a ladder diagram. The fact is there are millions of foreign engineers and programmers trained on the limited number of automation control systems. Also known as a *Fool's Paradise.*

security category: The characterization of information or an information system based on an assessment of the potential impact that a loss of confidentiality, integrity, or availability of such information or information system would have on organizational operations, organizational assets, or individuals. SOURCE: FIPS 200; FIPS 199; SP 800-18

Security Concept of Operations (Security CONOP): A security-focused description of an information system, its operational policies, classes of users, interactions between the system and its users, and the system's contribution to the operational mission. SOURCE: CNSSI-4009

Security Content Automation Protocol (SCAP): A method for using specific standardized testing methods to enable automated vulnerability management, measurement, and policy compliance evaluation against a standardized set of security requirements. SOURCE: CNSSI-4009

security control: The management, operational, and technical controls used to protect against an unauthorized effort to adversely affect the confidentiality, integrity, and availability of an information system or its information.

security control assessment: The testing and/or evaluation of the management, operational, and technical security controls in an information system to determine the extent to which the controls are implemented correctly, operating as intended, and producing the desired outcome with respect to meeting the security requirements for the system. SOURCE: SP 800-37; SP 800-53; SP 800-53A

security control assessor: The individual, group, or organization responsible for conducting a security control assessment. SOURCE: SP 800-37; SP 800-53A

security control enhancements: Statements of security capability to (1) build in additional, but related,

functionality to a basic control; and/or (2) increase the strength of a basic control. SOURCE: SP 800-53; SP 800-53A; SP 800-18; CNSSI-4009

security control inheritance: A situation in which an information system or application receives protection from security controls (or portions of security controls) that are developed, implemented, assessed, authorized, and monitored by entities other than those responsible for the system or application; entities either internal or external to the organization where the system or application resides. SOURCE: SP 800-37; SP 800-53; SP 800-53A

security Controls Baseline: The set of minimum security controls defined for a low-impact, moderate-impact, or high-impact information system. One of the sets of minimum security controls defined for federal information systems in NIST Special Publication 800-53 and CNSS Instruction 1253. SOURCE: SP 800-53; SP 800-53A; FIPS 200

security countermeasures: Actions, devices, procedures and techniques to reduce security risks.

security domain: A set of subjects, their information objects, and a common security policy. A collection of entities to which applies a single security policy executed by a single authority. A domain that implements a security policy and is administered by a single authority. SOURCE: SP 800-27; SP 800-37; SP 800-53; CNSSI-4009; FIPS 188

security engineering: An interdisciplinary approach and means to enable the realization of secure systems. It focuses on defining customer needs, security protection requirements, and required functionality early in the systems development life cycle, documenting requirements, and then proceeding with design, synthesis, and system validation while considering the complete problem. SOURCE: CNSSI-4009

security fault analysis (SFA): An assessment, usually performed on information system hardware, to determine the security properties of a device when hardware fault is encountered. SOURCE: CNSSI-4009

security features: Controls that protect against vulnerabilities— fire and water alarms, passwords, and other access protection—use of removable media for data storage, data validation controls, audit trails, uninterruptible power sources to protect against electrical outages, personnel screening, computer security awareness training of users, and so forth.

Security Features Users Guide (SFUG): Guide or manual explaining how the security mechanisms in a specific system work. SOURCE: CNSSI-4009

security filter: A secure subsystem of a building control system that enforces security policy on the data passing through it. SOURCE: CNSSI-4009

security impact analysis: The analysis conducted by an organizational official to determine the extent to which changes to the information system have affected the security state of the system. SOURCE: SP 800-53; SP 800-53A; SP 800-37; CNSSI-4009

security infraction: Failure to follow applicable laws, regulations, policies and procedures pertaining to the protection of data and computer resources. Infraction and violation are used interchangeably.

security kernel: Hardware, firmware, and software elements of a trusted computing base implementing the reference monitor concept. Security kernel must mediate all accesses, be protected from modification, and be verifiable as correct. SOURCE: CNSSI-4009

security label: The means used to associate a set of security attributes with a specific information object as part of the data structure for that object. A marking bound to a resource (which may be a data unit) that names or designates the security attributes of that resource. Information that represents or designates the value of one or more security relevant-attributes (e.g., classification) of a system resource. SOURCE: SP 800-53; FIPS 188; CNSSI-4009

security level: A hierarchical indicator of the degree of sensitivity to a certain threat. It implies, according to the security policy being enforced, a specific level of protection. SOURCE: FIPS 188

security markings: Human-readable indicators applied to a document, storage media, or hardware component to designate security classification, categorization, and/or handling restrictions applicable to the information contained therein. For intelligence information, these could include compartment and subcompartment indicators and handling restrictions. SOURCE: CNSSI-4009

security net control station: Management system overseeing and controlling implementation of network security policy. SOURCE: CNSSI-4009

security perimeter: A physical or logical boundary that is defined for a system, domain, or enclave, within which a particular security policy or security architecture is applied. SOURCE: CNSSI-4009

security plan: Formal document that provides an overview of the security requirements for the building control system and describes the security controls in place or planned for meeting those requirements. SOURCE: SP 800-53; SP 800-53A; SP 800-37; SP 800-18

security policy: The set of laws, rules, and practices that regulate how an organization manages, protects, and distributes sensitive information. Security policies define the objectives and constraints for the security program. Policies are created at several levels, ranging from organization or corporate policy to specific operational constraints (e.g., remote access). In general, policies provide answers to the questions "what" and "why" without dealing with "how." Policies are normally stated in terms that are technology-independent. SOURCE: SP 800-27

security posture: The security status of an enterprise's networks, information, and systems based on IA resources (e.g., people, hardware, software, policies) and capabilities in place to manage the defense of the enterprise and to react as the situation changes. SOURCE: CNSSI-4009

security program manager: In cybersecurity work, a person that manages BCS security (e.g., SCADA security) to include strategic, personnel, infrastructure, policy enforcement, emergency planning, security awareness, and other resources (e.g., the role of a Chief Information Security Officer).

security program plan: Formal document that provides an overview of the security requirements for an organization-wide information security program and describes the program management security controls and common security controls in place or planned for meeting those requirements. SOURCE: CNSSI-4009

security range: Highest and lowest security levels that are permitted in or on an information system, system component, subsystem, or network. SOURCE: CNSSI-4009

security-relevant change: Any change to a system's configuration, environment, information content, functionality, or users that has the potential to change the risk imposed upon its continued operations. SOURCE: CNSSI-4009

security-relevant event: An occurrence (e.g., an auditable event or flag) considered to have potential security implications to the system or its environment that may require further action (noting, investigating, or reacting). SOURCE: CNSSI-4009

security-relevant information: Any information within the information system that can potentially impact the operation of security functions in a manner that could result in failure to enforce the system security policy or maintain isolation of code and data. SOURCE: SP 800-53

security requirements: Requirements levied on an information system that are derived from applicable laws, Executive Orders, directives, policies, standards, instructions, regulations, or procedures, or organizational mission/business case needs to ensure the confidentiality, integrity, and availability of the information being processed, stored, or transmitted. SOURCE: FIPS 200; SP 800-53; SP 800-53A; SP 800-37; CNSSI-4009

security requirements traceability matrix (SRTM): Matrix that captures all security requirements linked to potential risks and addresses all applicable C&A requirements. It is a correlation statement of a building control system's security features and compliance methods for each security requirement. SOURCE: CNSSI-4009

security specification: A detailed description of the security requirements and specifications necessary to protect an ICS or SCADA installation. SOURCE: CNSSI-4009

security strength: A measure of the computational complexity associated with recovering certain secret and/or security-critical information concerning a given cryptographic algorithm from known data (e.g., plaintext/ciphertext pairs for a given encryption algorithm). A number associated with the amount of work (that is, the number of operations) that is required to break a cryptographic algorithm or system. Sometimes referred to as a *security level*. SOURCE: FIPS 186; SP 800-108

security target: Common Criteria specification that represents a set of security requirements to be used as the basis of an evaluation of an identified Target of Evaluation (TOE). SOURCE: CNSSI-4009

Security Test & Evaluation (ST&E): Examination and analysis of the safeguards required to protect an information system, as they have been applied in an operational environment, to determine the security posture of that system. SOURCE: CNSSI-4009

segment: A "single" section of a network with a limited number of locally-powered devices (typically 64 devices) that contains no repeaters or routers. There is generally a limit on the number of devices on a segment, and this limit is dependent on the topology/media and device type. For example, a TP/FT-10 segment with locally powered devices is limited to 64 devices, and a BACnet MS/TP segment is limited to 32 devices. SOURCE: UFGS 25 10 10

self-diagnostics: The checking by electronic means of the operability of a device.

Senior Agency Information Security Officer (SAISO): Official responsible for carrying out the chief information officer responsibilities under the Federal Information Security Management Act (FISMA) and serving as the chief information officer's primary liaison to the agency's authorizing officials, information system owners, and information system security officers. SOURCE: SP 800-53; SP 800-53A; SP 800-37; SP 800-60; FIPS 200; CNSSI-4009; 44 U.S.C., Sec. 3544

sensitive information: Information that requires a degree of protection due to its nature, magnitude of loss, or harm that could result from inadvertent or deliberate disclosure, modification, or destruction. This includes information that is:

- Mission critical (i.e., loss or harm would be such that an office could not perform essential functions).

- Should not be disclosed under the Freedom of Information Act, such as proprietary data and economic forecasts. Proprietary data includes trade secrets, commercial, or financial data obtained in the course of Government business, from or relating to a person or persons outside the government, not generally available to the public, and which is privileged, would cause competitive harm if released, or impair the ability of the government to obtain data in the future.

- Complies with OMB Circular A-127 Financial Management Systems.

- Complies with the Privacy Act of 1974. Data, which pertains to a specific individual by name, Social Security number or by some other identifying means, and is part of a system of records as defined in the Privacy Act of 1974.

- Classified.

sensor: A device that produces a voltage or current output that is representative of some physical property being measured (e.g., speed, temperature, airflow). There are many types of sensors:

- point sensors

- averaging sensors

- outside air temperature sensor

- carbon dioxide sensor

- humidity sensor

- differential pressure sensor

sensor analytics: Statistical analysis of data provided by sensors with the goal of detecting anomalies.

sensor interface module (SIM): Interface between sensors such as occupancy sensors and the BCS network. Typically enables each sensor to be independently configured.

sensory malware: Malware designed to hijack data collected surreptitiously from sensors on a networked device such as opportunistic images from a smartphone's camera, accelerometer, and geolocation information for reconnaissance purposes.

separation of duties (SOD): The principle of splitting privileges among multiple individuals or systems.

sequence of operation: Describes how building equipment is designed to be operated and is unique to each building. It describes proper startup, operation, and shutdown procedures for that particular facility only.

serial data transfer: Transmitting data one bit at a time.

Serial Storage Architecture (SSA): A high-speed method of connecting disk, tape, and CD-ROM drives, printers, scanners, and other devices to a computer.

server: A computer or intelligent electronic device that stores application and data files for all workstations on a BCS network; also referred to as a *file server*.

server crash: A server crash (or system crash) is when a computer program (such as a software application or an operating system) stops functioning properly. Failure is immediate. If the program is a critical part of the operating system, the entire computer may crash, often resulting in a kernel panic or fatal building control system error, or an unstable network. The process of debugging a crash is connecting the actual cause of the crash with the code that started the chain of events.

server hangs: A hang (also referred to as *freeze*) occurs when either a computer program or building control system ceases to respond to inputs. A typical example is a graphical user interface that no longer responds to the user's keyboard or mouse. A *hang* differs from a *server crash*, in which the failure is immediate and unrelated to the responsiveness of inputs.

service: A BACnet Service. A defined method for sending a specific type of data between devices. Services are always defined in a Client-Server manner, with a Client initiating a Service request and a Server executing the Service. Some examples are ReadProperty (a client requests a data value from a server), WriteProperty (a client writes a data value to a server), and CreateObject (a client requests that a server create a new object within the server device).

services: Software applications that facilitate communications to other applications or devices. Services are typically associated to a specific port.

service pin: A hardware push-button on a device (or initiated via software) that causes the device to broadcast a message containing its Node ID and Program ID. This broadcast can also be initiated via software.

servo controls: Servos are controlled by sending a pulse of variable width.

servo motor: A motor coupled to a sensor for position feedback that allows for precise control of angular position, velocity and acceleration.

servo system: A control system that converts a small mechanical motion into one requiring much greater power.

servo value: An actuated valve whose position is controlled using a servo actuator.

servo valve: Valves that transform a changing analog or digital input signal into a smooth set of steps of movement in a hydraulic cylinder.

session hijacking attack: Taking over a session that someone else established.

setpoint: Also set point, is the desired or target value for an essential variable of a building control system to describe a standard configuration or norm for the building control system. For example, a boiler might have a *temperature setpoint*, which is the temperature the boiler control system aims to maintain.

setpoint override: Most setpoint loads in a heating system operate year round at temperatures above the reset operating temperature. When the setpoint load requires heat, this overrides the Warm Weather Shut Down and Reset temperature of the control. This function allows the heating system to operate only at the temperature required to satisfy the current load.

shadow password file: A building control system file in which encrypted user passwords are stored so that they aren't available to people who try to break into the building control system.

shadow warfare: Another term for *cyber-warfare*.

shaft encoder: A device that converts the angular position or motion of a shaft or axle to an analog or digital code.

shafting hack: Buildings have maintenance shafts for passage of pipes and ducts between floors. Climbing inside these shafts is known as *shafting*. This is similar to *buildering*, which is done on the outsides of buildings. A dangerous variant of shafting involves entering elevator shafts, either to ride on the top of the elevators, or to explore the shaft itself. This activity is sometimes called *elevator surfing*. Extremely dangerous, and thanks to Darwin theory, extremely rare.

Shamoon attack: An extremely destructive virus consisting of three components: a dropper, a wiper, and a reporter module.

- **Dropper** is responsible for creating the required files on the building control system, registering a service called TrkSvr in order to start itself with Windows. It also copies itself to accessible network shares and executes itself remotely.

- **Wiper** is only activated when a hardcoded configuration date has been passed. This enables a coordinated, "time bomb" scenario. The module drops a legitimate and digitally signed device driver that provides low-level disk access from user space. The malware collects file names and starts overwriting them with a JPEG image or blocks of random data. **Disttrack** finishes off the computer by wiping the master boot record with the same data.

- **Reporter** is responsible for sending back information to the control server. It reports the domain name, IP address, and number of files overwritten.

shielded enclosure: Room or container designed to attenuate electromagnetic radiation, acoustic signals, or emanations. SOURCE: CNSSI-4009

shielding: Reducing the electromagnetic field in a space by blocking the field with barriers made of conductive or magnetic materials.

Shodan: Shodan is a search engine that lets a user find specific types of computers (routers, servers, etc.) connected to the Internet using a variety of filters. This can be information about the server software, what options the service supports, a welcome message or anything else that the client can find out before interacting with the server. Shodan searches the Internet for publicly accessible devices, concentrating on SCADA systems. If your building control system is listed on Shodan, it probably can be hacked.

SHODANDiggity: Easy interface to Shodan search engine.

SHODAN Hacking Database (SHDB): A dictionary file containing queries used to target Building Control Systems, Industrial Control Systems and SCADA systems.

shadow file processing: An approach to data backup in which real-time duplicates of critical files are maintained at a remote processing site. Similar term: *Remote Mirroring*

Short Message Service (SMS): A text messaging service component of mobile communication. Standardized communications protocols are used to allow fixed line or mobile phone devices to exchange short text messages of up to 160 characters.

short-term exposure limit (STEL): Exposure to hazardous gas; usually monitored over 15-minute periods.

shoulder surfing: Using direct observation technique, such as looking over one's shoulder, to obtain their personal access information such as passwords, PINs, and security codes.

shunt: A device that allows electrical current to pass around another point in the circuit.

sideloading attack: A term used mostly on the Internet, similar to *upload* and *download*, but in reference to the process of transferring data between two local devices, in particular between a computer and a mobile device such as a mobile phone, smartphone, PDA, tablet, portable media player, or e-reader.

signature: A recognizable, distinguishing pattern.

signature certificate: A public key certificate that contains a public key intended for verifying digital signatures rather than encrypting data or performing any other cryptographic functions. SOURCE: SP 800-32; CNSSI-4009

Simple Network Management Protocol (SNMP): An agreed upon framework by which building control system network devices can share information about their status.

simulation test: A test of recovery procedures under conditions approximating a specific cyber-attack scenario. This may involve designated units of the organization actually ceasing normal operations while exercising their procedures.

single-hop problem: The security risks resulting from a mobile software agent moving from its home platform to another platform. SOURCE: SP 800-19

single sign-on: Authentication to enable a user to authenticate once and gain access to multiple software systems. A dangerous practice, to be discouraged.

single point of failure (SPOF): A part of a building control system that, if it fails, will stop the entire building control system from functioning properly.

sinkholing: The redirection of traffic from its original destination to one specified by the sinkhole owners. The altered destination is known as the *sinkhole*. Sinkholes can be good or bad. A *botnet sinkhole* is used to gather information about a particular botnet (that's good).

situational awareness: The perception of an enterprise's security posture and its threat environment within a volume of time and space; the comprehension/meaning of both taken together (risk); and the projection of their status into the near future. SOURCE: CNSSI-4009

skills inventory: A listing of employees that lists their skills that apply to cyber-recovery.

skimming: The unauthorized use of a reader to read tags without the authorization or knowledge of the tag's owner or the individual in possession of the tag. SOURCE: SP 800-98

Small Computer System Interface (SCSI): An interface that serves as an expansion bus that can be used to connect hard disk drives, tape drives, and other hardware components.

smart card: A credit card-sized plastic card with embedded integrated circuits that can store, process, and communicate information. SOURCE: CNSSI-4009

smart grid: An electrical grid that includes a variety of operational and energy measures including *smart meters, smart appliances, renewable energy resources*, and *energy efficiency resources*.

smart meter: An electronic device that records consumption of electric energy in intervals of an hour or less and communicates that information at least daily back to the utility for monitoring and billing. Smart meters differ from traditional Automatic Meter Reading (AMR) in that smart meters enable two-way communications with the meter and the central system. Unlike home energy monitors, smart meters can gather data for remote reporting. Although "smart meter" often refers to an electricity meter, it also may mean a device measuring natural gas or water consumption. Totally hackable. They are also referred to as *interval* or *time-of-use meters*. SOURCE: Wikipedia

SmartPhone Ad hoc Networks (SPANs): Leverage the existing hardware (primarily Bluetooth and Wi-Fi) in commercially available smartphones to create peer-to-peer networks without relying on cellular carrier networks, wireless access points, or traditional network infrastructure.

smart thermostat: A thermostat with Wi-Fi capability designed to improve energy efficiency and control the temperature from a computer, tablet, or smartphone. Some smart thermostats are voice-enabled (voice command). That means that you can ask it to make temperature adjustments totally hands-free. Voice-enabled thermostats can be retasked to eavesdrop on conversations and are a significant security risk. *Do not use these in secure areas.*

smishing attack: A form of phishing attack that uses a cell phone text message (SMS) to lure a target to a web site or prompting the target to a call a telephone number in an attempt to persuade target to reveal credit card information such as pin number. Often a smishing target is informed he will be charged for something unless he clicks on a link and cancels it.

Smurf attack: A cyber-attack that spoofs the target address and sends a ping to the broadcast address for a remote network, which results in a large amount of ping replies being sent to the target.

snarf: To grab, especially to grab a large document or file for the purpose of using it with or without the author's permission. To fetch a file or set of files across a network.

Sneakernet: An informal term describing the transfer of electronic information, especially computer files, by physically moving removable media such as magnetic tape, floppy disks, compact discs, USB flash drives (thumb drives, USB stick) or external hard drives from one computer to another, usually in lieu of transmitting the information over a computer network.

sniffer: Is a computer program or piece of computer hardware that can intercept and log traffic that passes over a digital network or part of a network. As data streams flow across the network, the sniffer captures each packet and, if needed, decodes the packet's raw data, showing the values of various fields in the packet, and analyzes its content according to the appropriate RFC or other specifications.

snoop server: A server that uses a packet sniffer program to capture network traffic for analysis. Used to identify security risks and/or to monitor employees' activities (such as web sites visited), a snoop program puts network interfaces into promiscuous mode.

snort: Popular open source intrusion detection system software.

snort and dragon sensors: Signature-matching intrusion detection applications that report alerts and provide information on source and destination IP, and port, and which rule or signature was violated.

Social-Engineer Toolkit (SET): A unique tool that attacks the human element rather than the building control system. It has features that let you send e-mails and Java applets containing cyber-attack code. It is to be used very carefully and only for "white hat" purposes.

social engineering attack: Social engineering is the art and science of getting people to do something you want them to do that they might not do in the normal course of action. Instead of collecting information by technical means, intruders might also apply methods of social engineering such as impersonating individuals on the telephone, or using other persuasive means (e.g., tricking, convincing, inducing, enticing, provoking) to encourage someone to disclose information. Attackers look for information about who the target does business with, both suppliers and customers and they are particularly interested in IT support. They gather this information to better understand roles and responsibilities. They use this information to pose as someone from one of these companies. Attackers look for information such as birthdays, who was recently promoted or who just had a baby. Hackers do not discount any information they uncover. They will use bad relationships between IT department and other offices as a wedge to gain information.

social spam: Unwanted spam content appearing on social networks and any web site with user-generated content (comments, chat, etc.). It can appear in many forms, including bulk messages, profanity, insults, hate speech, malicious links, fraudulent reviews, fake friends, and personally identifiable information.

SOPHIA: Passive network monitoring tool developed by Idaho National Labs.

soft access point (Soft AP): A Soft Access Point can be set up on a Wi-Fi adapter without the need of a physical Wi-Fi router. With Windows 7 virtual Wi-Fi capabilities and Intel My Wi-Fi technology, one can easily set up a Soft AP on a machine. Once up and running, one can share the network access available on a machine with other Wi-Fi users that will connect to the soft AP. If any employee sets up a Soft AP on their machine inside the corporate premises and shares the corporate network through it, then this Soft AP behaves as Rogue Access Point.

soft zero: When a soft zero is performed on a gas detector or monitor (i.e., the *autozero* on startup) any adjustments will only remain in place while the instrument remains switched on.

software assurance: The level of confidence that software is free from vulnerabilities, either intentionally designed into the software or accidentally inserted at any time during its lifecycle, and that the software functions in the intended manner. SOURCE: CNSSI-4009

software-based fault isolation: A method of isolating application modules into distinct fault domains enforced by software. The technique allows untrusted programs written in an unsafe language, such as C, to be executed safely within the single virtual address space of an application. Untrusted machine interpretable code modules are transformed so that all memory accesses are confined to code and data segments within their fault domain. Access to system resources can also be controlled through a unique identifier associated with each domain. SOURCE: SP 800-19

software radio: A radio used by hobbyists that performs the signal processing using software instead of in the hardware. Because software can be replaced, the same device can be used for many different applications. See *Universal Software Radio Peripheral (USRP)*.

solenoid valve: A valve actuated by an electric coil. A solenoid valve typically has two states: open and closed.

solid-state relay (SSR): Provide a high degree of reliability, long life and reduced electromagnetic interference (EMI), together with fast response and high vibration resistance, as compared to an electromechanical relay (EMR). All the advantages of solid state circuitry, including consistency of operation and a typically longer usable lifetime because it has no moving parts to wear out or arcing contacts to deteriorate, which are primary causes of failure of an electromechanical relay.

space cyber: Cyber activity unique to the design and operation of space-based systems and technologies.

spam: The abuse of electronic messaging systems to indiscriminately send unsolicited bulk messages. SOURCE: SP 800-53 Malicious spam (*malspam*) sometimes contains a Word document designed to infect a Windows computer with malware.

spam blogs: Spam blogs are blogs created solely for commercial promotion and the passage of link authority to target sites. Often these "splogs" are designed in a misleading manner that gives the effect of a legitimate web site, but upon close inspection is often written using spinning software or very poorly written and barely readable content. They are similar in nature to *link farms*.

spamdexing: The deliberate manipulation of search engine indexes. It involves a number of methods, such as repeating unrelated phrases, to manipulate the relevance or prominence of resources indexed in a manner inconsistent with the purpose of the indexing system. Also known as *search engine spam, search engine poisoning, search spam*, or *web spam*.

spark detection: Production machines, dryers, mills, sanders, ovens, grinders, pelletizers, buffers, furnaces, and shot blasting are all sources of sparks that can cause fires and explosions in dust collection systems. IR sensors will detect sparks and signal the control console to immediately activate the extinguishing devices, activate the spark alarm lamp and horn, and record the event. A fine mist of water spray is produced to extinguish the sparks or embers. The control console confirms and records the water flow indication at the extinguishing devices.

spear phishing attack: Phishing attempts directed at specific individuals or companies with the sole purpose of obtaining unauthorized access to victim's sensitive data such as network access credentials. Attackers may initially gather personal information about their target to increase the probability of success. This technique is, by far, the most successful on the Internet today, accounting for 91% of cyber-attacks.

special access program: A program established for a specific class of classified information that imposes safeguarding and access requirements that exceed those normally required for information at the same classification level. SOURCE: SP 800-53; CNSSI-4009

special access program facility (SAPF): Facility accredited by an appropriate agency in accordance with ICD 706 in which SAP information may be processed. SOURCE: CNSSI-4009

special character: Any non-alphanumeric character that can be rendered on a standard American-English keyboard. Use of a specific special character may be application-dependent. SOURCE: CNSSI-4009 The list of special characters follows:

$$` ~ ! @ \# \$ \% \wedge \& * (\ ) \_ + | \} \{ " : ? > < [\ ] \backslash ; ' , . / : =$$

spike mic: (1) A gadget that can fire a special dart with built-in microphone and secretly listen to nearby conversations. The launcher acts as a receiver and has a listening range of up to 50 feet. (2) A spike that goes physically through the wall into the next room mechanically coupled to the diaphragm of a microphone on the other side of the wall.

spillage: Security incident that results in the transfer of classified or CUI information onto an information system not accredited (i.e., authorized) for the appropriate security level. SOURCE: CNSSI-4009

spim: Unsolicited instant messages from someone you don't know attempting to sell you something.

spindle: Mechanism inside a hard drive that moves the heads into place; the axle on which a disk turns.

split-horizon: A method of preventing routing loops in distance-vector routing protocols by prohibiting a router from advertising a route back onto the interface from which it was learned.

spoofing: (1) Faking the sending address (IP, Caller ID, GPS, e-mail address) of a transmission to gain illegal (unauthorized) entry into a secure building control system. (2) Spoofing can also refer to legitimate copyright holders placing distorted or unlistenable versions of their works on file-sharing networks. SOURCE: SP 800-48

spoofing attack: Generation of outbound network traffic pretending to be from somewhere else, typically used in a denial-of-service attack. See *Masquerading Attack*.

spyware: Software that aims to gather information about a person or organization without their knowledge and that may send such information to another entity without the consumer's consent, or that asserts control over a computer without the consumer's knowledge. Spyware is mostly classified into four types: system monitors, Trojan horse, adware, and tracking cookies. Spyware is mostly used for the purposes of tracking and storing Internet users' movements on the web and serving up pop-up ads to Internet users. Whenever spyware is used for malicious purposes, its presence is typically hidden from the user and can be difficult to detect. Some spyware, such as *keyloggers*, may be installed by the owner of a shared, corporate, or public computer intentionally in order to monitor users. While the term spyware suggests software that monitors a user's computing, the functions of spyware can extend beyond simple monitoring. Spyware can collect almost any type of data, including personal information like Internet surfing habits, user logins, and bank or credit account information. Spyware can also interfere with user control of a computer by installing additional software or redirecting web browsers. Some spyware can change computer settings, which can result in slow Internet connection speeds, unauthorized changes in browser settings, or changes to software settings. Spyware does not necessarily spread in the same way as a virus or worm because infected systems generally do not attempt to transmit or copy the software to other computers.

▦ **Note**   Some spyware cannot be completely removed.

These common spyware programs illustrate the diversity of behaviors found in these attacks. Note that as with computer viruses, researchers give names to spyware programs, which may not be used by their creators. Programs may be grouped into "families" based not on shared program code, but on common behaviors, or by "following the money" of apparent financial or business connections. Programs that are frequently installed together may be described as parts of the same spyware package, even if they function separately.

- **CoolWebSearch**, a group of programs, takes advantage of Internet Explorer vulnerabilities. The package directs traffic to advertisements on web sites including coolwebsearch.com. It displays pop-up ads, rewrites search engine results, and alters the infected computer host's file to direct DNS lookups to these sites.

- **FinFisher**, sometimes called **FinSpy** is a high-end surveillance suite sold to law enforcement and intelligence agencies. Support services such as training and technology updates are part of the package.

- **HuntBar**, aka **WinTools** or **Adware**. Websearch was installed by an ActiveX drive-by download at affiliate web sites, or by advertisements displayed by other spyware programs—an example of how spyware can install more spyware. These programs add toolbars to IE, track aggregate browsing behavior, redirect affiliate references, and display advertisements.

- **Internet Optimizer**, also known as **DyFuCa**, redirects Internet Explorer error pages to advertising. When users follow a broken link or enter an erroneous URL, they see a page of advertisements. However, because password-protected web sites use the same mechanism as HTTP errors, Internet Optimizer makes it impossible for the user to access password-protected sites.

- Spyware such as **Look2Me** hides inside system-critical processes and start up even in safe mode. With no process to terminate they are harder to detect and remove, which is a combination of both spyware and a rootkit. Rootkit technology is also seeing increasing use, as newer spyware programs also have specific countermeasures against well-known anti-malware products and may prevent them from running or being installed, or even uninstall them.

- **Movieland**, also known as **Moviepass.tv** and **Popcorn.net**, is a movie download service that has been the subject of thousands of complaints to the Federal Trade Commission and other agencies.

- **WeatherStudio** has a plug-in that displays a window-panel near the bottom of a browser window.

- **Zango** transmits detailed information to advertisers about the web sites that users visit. It also alters HTTP requests for affiliate advertisements linked from a web site, so that the advertisements make unearned profit for the company. It opens pop-up ads that cover over the web sites of competing companies.

- **Zlob Trojan**, or just Zlob, downloads itself to a computer via an ActiveX codec and reports information back to Control Server. Some information can be the search-history, the web sites visited, and even keystrokes. More recently, Zlob has been known to hijack routers set to defaults.

spyware attack: Software that is secretly or surreptitiously installed onto a BCS to gather information; a type of malicious code that monitors or spies on its victims. It usually remains in hiding. SOURCE: SP 800-53; CNSSI-4009

spy-phishing: Defined as "crimeware," spy-phishing capitalizes on the trend of "blended threats." It borrows techniques from both phishing and spyware. The downloaded applications sit silently on the user's system until the targeted URL is visited wherein it activates, sending information to the malicious third party. Through the use of spyware and other Trojans, spy-phishing attempts to prolong the initial phishing attacks beyond the point at which the phishing site is available.

SQL injection attack: A type of *input validation attack* where SQL code is inserted into database-driven application queries to manipulate the database.

stack smashing attack: A cyber-attack using a buffer overflow to trick a computer into executing arbitrary code.

Stages of a cyber-attack:

- **Survey**: Investigating and analyzing available information about the target in order to identify potential vulnerabilities. Attackers will use any means available to find technical, procedural, or physical vulnerabilities that they can attempt to exploit. They will use open source information such as LinkedIn and Facebook, domain name management/search services, and social media. They will employ commodity toolkits and techniques, and standard network scanning tools to collect and assess any information about your organization's computers, security systems, and personnel.

- **Delivery**: Getting to the point in a building control system where a vulnerability can be exploited. The attacker will look to get into a position where they can exploit a vulnerability that they have identified, or they think could potentially exist. The crucial decision for the attacker is to select the best delivery path for the malicious software or commands that will enable them to breach your defenses. Examples include the following:

  - attempting to access an organization's online services

  - sending an e-mail containing a link to a malicious web site or an attachment which contains malicious code

  - giving an infected USB stick away at a trade fair

  - creating a false web site in the hope that a user will visit

- **Breach**: Exploiting the vulnerability/vulnerabilities to gain some form of unauthorized access. The damage will depend on the nature of the vulnerability and the exploitation method. Having done this, the attacker could pretend to be the victim and use their legitimate access rights to gain access to other systems and information. It may allow them to:

  - make changes that affect the building control system's operation

  - gain access to online accounts

  - achieve full control of a user's computer, tablet or smartphone

- **Affect**: Carrying out activities within a building control system that achieve the attacker's goal. Depending on their motivation, the attacker may seek to explore your building control systems, expand their access, and establish a persistent presence (a process sometimes called *consolidation*). Taking over a user's account usually guarantees a persistent presence. Taking over an administrator's account is an attacker's Holy Grail. With administration access to just one building control system, they can try to install automated scanning tools to discover more about your networks and take control of more systems. When doing this they will take great care not to trigger the building control system's monitoring processes and they may even disable them for a time.

stand-alone processing: Processing, typically on a PC or mid-range computer, which does not require any communication link with a mainframe or other processor.

standby power: The electric power consumed by electronic devices and electrical appliances while they are switched off (but are designed to continue to draw some power) or in a standby mode. Also called *vampire power, vampire draw, phantom load*, or *leaking electricity*.

standard object/property/service: Objects, properties, or services that are standard objects, properties, or services enumerated and defined in ASHRAE 135. Clause 23 of ASHRAE 135 defines methods to extend ASHRAE 135 to non-standard or proprietary information. Standard BACnet objects/properties/services specifically exclude any vendor-specific extensions.

Standard Configuration Property Type (SCPT): Standard format for configuration properties.

Standard Network Variable Type (SNVT): Standard format type to define data information transmitted and received by individual nodes.

Statistical Process Control (SPC): The use of statistical techniques to control the quality of a product or process.

stealth strategy attack: Some viruses try to avoid detection by killing the tasks associated with antivirus software before it can detect them (for example, **Conficker** worm, also known as **Downup**, **Downadup**, and **Kido**).

stealware: A type of malware that covertly transfers money or data to a third party. Stealware uses an HTTP cookie to redirect the commission ordinarily earned by a web site for referring users to another site.

steganography: The art and science of communicating in a way that hides the existence of the communication. For example, a child pornography image can be hidden inside another graphic image file, audio file, or other file format. A real-world example is "invisible" ink. SOURCE: SP 800-72; SP 800-101

storage area network (SAN): A network infrastructure of shared multi-host storage, linking all storage devices as well as interconnecting remote sites.

strength of mechanism (SoM): A scale for measuring the relative strength of a security mechanism. SOURCE: CNSSI-4009

striped core: A network architecture in which user data traversing a core IP network is decrypted, filtered and re-encrypted one or more times. Note: The decryption, filtering, and re-encryption are performed within a "Red gateway"; consequently, the core is "striped" because the data path is alternately black, red, and black. SOURCE: CNSSI-4009

striping: A method of data storage in which a unit of data is distributed and stored across several hard disks, which improves access speed, but does not provide redundancy.

strong authentication: The requirement to use multiple factors for authentication and advanced technology, such as dynamic passwords or digital certificates, to verify an entity's identity. SOURCE: CNSSI-4009

structured walkthrough test: Team members walk through the cyber-attack recovery plan to identify and correct weaknesses.

Stuxnet attack: This virus allegedly targeted the PLCs and modified valve settings. Closing valves at certain points in time would lead to an increase of pressure that could damage equipment. The later version of the threat focused on the PLCs manipulating the spinning frequency of rotating motors. By speeding motors up and slowing them down repeatedly, the output quality could be spoiled and equipment could be damaged. To hide its activity, Stuxnet allegedly executed slightly different infection routines depending on the security software installed on the target network. To avoid detection by personnel monitoring the plant, Stuxnet allegedly recorded measurement readings during normal operation and played those back in a loop.

subject: The person whose identity is bound to a particular credential. SOURCE: SP 800-63

subnet: A logical grouping of up to 127 nodes, where the logical grouping is defined by node addressing. Each subnet is assigned a number that is unique within the domain.

subscriber identity module (SIM): A removable smart card containing a cell phone's subscription information and phone book.

subscription: Contract commitment providing an organization with the right to utilize a vendor recovery facility for recovery of their mainframe processing capability.

superencryption: Process of encrypting encrypted information. Occurs when a message, encrypted offline, is transmitted over a secured, online circuit, or when information encrypted by the originator is multiplexed onto a communications trunk, which is then bulk encrypted. SOURCE: CNSSI-4009

supercookie: A highly persistent tracking cookie placed on a user's computer.

super-user: A building control system account that has full building control system-wide administrative privileges. Most Unix machines have a log on account called *root*, which acts as the super-user.

supervisory controller: A controller implementing a combination of supervisory logic (global control strategies or optimization strategies), scheduling, alarming, event management, trending, web services, or network management. Note this is defined by use; many supervisory controllers have the capability to also directly control building equipment.

supervisory control and data acquisition (SCADA): A generic name for a computerized system that is capable of gathering and processing data and applying operational controls over long distances. Typical uses include power transmission and distribution and pipeline systems. SCADA was designed for the unique communication challenges (delays, data integrity, etc.) posed by the various media that must be used, such as phone lines, microwave, and satellite. Usually shared rather than dedicated. SOURCE: SP 800-82 Networks or systems generally used for industrial controls or to manage infrastructure such as pipelines and power systems. SOURCE: CNSSI-4009

supervisory gateway: A device that is both a supervisory controller and a gateway, such as a Niagara Framework Supervisory Gateway.

supplementation (assessment procedures): The process of adding assessment procedures or assessment details to assessment procedures in order to adequately meet the organization's risk management needs. SOURCE: SP 800-53A

supplementation (security controls): The process of adding security controls or control enhancements to a security control baseline from NIST Special Publication 800-53 or CNSS Instruction 1253 in order to adequately meet the organization's risk management needs. SOURCE: SP 800-53A

supply chain: A system of organizations, people, activities, information and resources, for creating and moving products including product components and/or services from suppliers through to their customers. SOURCE: SP 800-53; CNSSI-4009

supply chain attack: Attacks that allow the adversary to utilize implants or other vulnerabilities inserted prior to installation in order to infiltrate data, or manipulate hardware, software, operating systems, peripherals (information technology products) or services at any point during the life cycle. For example, including a tiny microphone in millions of thermostats manufactured in a foreign country so when they are installed in sensitive rooms, they can be used to eavesdrop on conversations. SOURCE: CNSSI-4009

supply chain risk management: The process of identifying, analyzing, and assessing supply chain risk and accepting, avoiding, transferring or controlling it to an acceptable level considering associated costs and benefits of any actions taken.

surreptitious entry: Unauthorized entry in a manner that leaves no readily discernable evidence.

sustained mode: The measured transfer rate of a given device during normal operation.

switch: A network traffic-monitoring device that controls the flow of traffic between multiple network nodes.

switchgear: In an electric power system, switchgear is the combination of electrical disconnect switches, fuses or circuit breakers used to control, protect and isolate electrical equipment. One of the main functions of switchgear is to protect the system against faults including failures of components like load transformers, cable, and cable terminations. Used both to de-energize equipment to allow work to be done, and to clear faults downstream. Usually, personnel trained in medium-voltage switchgear who can manually operate breakers or switches can restore service to the building from an alternate power source. A switchgear design with manual transfer is recommended. This requires two feeds from the utility, allowing an opportunity to manually switch to the second source to easily restore service.

Sybil cyber-attack: A Sybil cyber-attack is the forging of multiple identities for malicious intent, named after "Sybil," the famous multiple personality disorder patient. A spammer may create multiple web sites at different domain names that all link to each other, such as fake blogs (known as *spam blogs*).

syllabary: List of individual letters, combination of letters, or syllables, with their equivalent code groups, used for spelling out words or proper names not present in the vocabulary of a code. A syllabary may also be a *spelling table*. SOURCE: CNSSI-4009

symmetric cryptography: A branch of cryptography in which a cryptographic system or algorithms use the same secret key (a shared secret key).

symmetric key: A cryptographic key that is used to perform both the cryptographic operation and its inverse, for example to encrypt plaintext and decrypt ciphertext, or create a message authentication code and to verify the code.

SYN Flood attack: A denial-of-service attack that sends a host more TCP SYN packets than the protocol can handle.

Synchronous Serial Interface (SSI): A widely used serial interface standard for industrial applications between a master controller and a sensor.

system: A generic term used for its brevity to mean either a major application or a general support system. A building control system is identified by logical boundaries drawn around the various processing communications, storage, and related resources. They must be under the same direct management control (not responsibility), perform essentially the same function, reside in the same environment, and have the same characteristics and security needs. A system does not have to be physically connected. SOURCE: CNSSI-4009

system administrator: A person who manages the technical aspects of a system. Individual responsible for the installation and maintenance of an information system, providing effective information system utilization, adequate security parameters, and sound implementation of established information assurance policies and procedures. SOURCE: SP 800-40; CNSSI-4009

system high mode: Information systems security mode of operation wherein each user, with direct or indirect access to the information system, its peripherals, remote terminals, or remote hosts, has all of the following: (1) valid security clearance for all information within an information system; (2) formal access approval and signed nondisclosure agreements for all the information stored and/or processed (including all compartments, subcompartments and/or special access programs); and (3) valid need-to-know for some of the information contained within the information system. SOURCE: CNSSI-4009

systems downtime: A planned interruption in building control system availability for scheduled building control system maintenance.

systems integrator: An individual or company that combines various components and programs into a functioning system, customized for a particular customer's needs.

system integrity: The attribute of a building control system when it performs its intended function in an unimpaired manner, free from deliberate or inadvertent unauthorized manipulation of the building control system.

system outage: An unplanned interruption in building control system availability as a result of computer hardware or software problems, or operational problems.

system profile: Detailed security description of the physical structure, equipment component, location, relationships, and general operating environment of an information system. SOURCE: CNSSI-4009

systems requirements planning: In cybersecurity work, a person that consults with customers to gather and evaluate functional requirements and translates these requirements into technical solutions; provides guidance to customers about applicability of building control systems to meet business needs.

systems security analysis: In cybersecurity work, a person that conducts the integration/testing, operations, and maintenance of building control systems security.

systems security architecture: In cybersecurity work, a person that develops building control system concepts and works on the capabilities phases of the building control systems development lifecycle; translates technology and environmental conditions (e.g., law and regulation) into building control system and security designs and processes.

System Security Plan (SSP): Formal document that provides an overview of the security requirements for the information system and describes the security controls in place or planned for meeting those requirements. The formal document prepared by the information system owner (or common security controls owner for inherited controls) that provides an overview of the security requirements for the system and describes the security controls in place or planned for meeting those requirements. The plan can also contain as supporting appendices or as references, other key security-related documents such as a risk assessment, privacy impact assessment, system interconnection agreements, contingency plan, security configurations, configuration management plan, and incident response plan. SOURCE: SP 800-37; SP 800-53; SP 800-53A; SP 800-18; FIPS 200; CNSSI-4009

CHAPTER 21

T

tabletop exercise: A discussion-based exercise where personnel with roles and responsibilities in a particular IT plan meet in a classroom setting or in breakout groups to validate the content of the plan by discussing their roles during an emergency and their responses to a particular emergency situation. A facilitator initiates the discussion by presenting a scenario and asking questions based on the scenario. SOURCE: SP 800-84

tactical edge: The platforms, sites, and personnel (U.S. military, allied, coalition partners, first responders) operating at lethal risk in a battle space or crisis environment characterized by (1) a dependence on information systems and connectivity for survival and mission success, (2) high threats to the operational readiness of both information systems and connectivity, and (3) users are fully engaged, highly stressed, and dependent on the availability, integrity, and transparency of their information systems. SOURCE: CNSSI-4009

tag: In computer language, a tag is simply a *variable*.

tag command queuing (TCQ): A feature introduced in the SCSI-2 specification that permits each initiator to issue commands accompanied by instructions for how the target should handle the command; the initiator can either request the command to be executed at the first available opportunity, in the order in which the command was received, or at a time deemed appropriate by the target.

tailoring: The process by which a security control baseline is modified based on (1) the application of scoping guidance; (2) the specification of compensating security controls, if needed; and (3) the specification of organization-defined parameters in the security controls via explicit assignment and selection statements. SOURCE: SP 800-37; SP 800-53; SP 800-53A; CNSSI-4009

tailored trustworthy space: A cyberspace environment that provides a user with confidence in its security, using automated mechanisms to ascertain security conditions and adjust the level of security based on the user's context and in the face of an evolving range of threats.

takeover hack: Cyber-attack of an automobile over the Internet to hijack its controls remotely such as the brakes and transmission of a vehicle on the road.

tampering attack: Tampering is a web-based cyber-attack where certain parameters in the URL are changed without the customer's knowledge; and when the customer keys in that URL, it looks and appears exactly the same. Tampering is basically done by hackers and criminals to steal the identity and obtain illegal access to information. SOURCE: CNSSI-4009

target: A SCSI device that performs an operation requested by an initiator.

technical tattoos (tech tats): The use of skin-mounted components and conductive paint to create circuitry in the form of a tattoo that has the capability to collect, store, send and receive data.

© Luis Ayala 2016
L. Ayala, *Cybersecurity Lexicon*, DOI 10.1007/978-1-4842-2068-9_21

technical controls: The security controls (i.e., safeguards or countermeasures) for an ICS that are primarily implemented and executed by the building control system through mechanisms contained in the hardware, software, or firmware components of the building control system. SOURCE: SP 800-53; SP 800-53A; SP 800-37; FIPS 200

technical security: A security discipline dedicated to detecting, neutralizing or exploiting a wide variety of hostile and foreign penetration technologies.

technical security controls: Security controls (i.e., safeguards or countermeasures) for an information system that are primarily implemented and executed by the information system through mechanisms contained in the hardware, software, or firmware components of the system. SOURCE: CNSSI-4009

technical threats: A disaster-causing event that may occur regardless of any human elements.

technical vulnerability information: Detailed description of a weakness to include the implementable steps (such as code) necessary to exploit that weakness. SOURCE: CNSSI-4009

telco: Abbreviation for a telecommunications company.

television hacks: Computer Security firm ReVuln claimed it could hack Samsung's newest televisions, access user settings, install malware on the TVs and any connected devices, and harvest the personal data stored on the TV. They could switch on the camera embedded in the TV and watch viewers watching the set. Google and Verizon have been reported developing cable TV boxes with built-in video cameras and motion sensors. If the camera detects two people on the couch, they might be delivered ads for a romantic movie, while a room full of children would see advertisements targeting children.

temperature limit switch: Intended to be used with industrial heating equipment to prevent excess temperature if the temperature-controlling equipment fails. It is preferable that temperature-limit switch action automatically shut down the heating system, and to do this, switch contacts that are closed during normal operation of the equipment are generally used. In the event of excess temperature, a manual action is required to restore the switch contacts.

temperature sensor: A sensor system that produces an electrical signal related to its temperature, and as a consequence, senses the temperature of its surrounding medium. Electronic temperature sensors are designed for use with electronic controllers in domestic or commercial heating and cooling systems. Wallplate temperature sensors provide a resistive output signal proportional to sensed room or space temperature.

TEMPEST: Refers to the investigation, study, and control of unintentional compromising emanations from telecommunications and automated information systems equipment. SOURCE: FIPS 140-2

TEMPEST zone: Designated area within a facility where equipment with appropriate TEMPEST characteristics (TEMPEST zone assignment) may be operated. SOURCE: CNSSI-4009

temporary operating procedures: Predetermined procedures, which streamline operations while maintaining an acceptable level of control and auditability during a cyber-physical attack.

terabyte: Approximately one trillion bytes; 1,024 gigabytes.

terminal block: A convenient way to connect wires to a single electrical connection point where a wire is held by the tightening of a screw. These blocks allow wires to be joined without the need to splice them, so wiring can easily be adjusted later if needed.

test & evaluation: In cybersecurity work, a person that develops and conducts tests of building control systems to evaluate compliance with specifications and requirements by applying principles and methods for cost-effective planning, evaluating, verifying, and validating of technical, functional, and performance characteristics (including interoperability) of building control systems or elements of building control systems incorporating information technology.

test plan: The recovery plans and procedures that are used in a building control systems test to ensure viability. A test plan is designed to exercise specific action tasks and procedures that would be encountered in a real disaster.

test scenarios: Are descriptions of the tests to be performed to check the effectiveness of the security features. They may include validation of password constraints, such as length and composition of the password, entry of erroneous data to check data validation controls, review of audit information produced by the building control system, review of contingency plans and risk analyses, and so forth.

tethering attack: Connecting one device to another. In the context of mobile phones and tablet computers, tethering allows sharing the Internet connection of the phone or tablet with other devices such as laptops. Connection of the phone or tablet with other devices can be done over wireless LAN (Wi-Fi), over Bluetooth or by physical connection using a cable, for example through USB. Also called a *personal hotspot.*

Tetris on the Green hack: Hackers turned MIT's Green Building into a giant, playable, and multi-color *Tetris* game using lights in building windows. A console allowed players to move, rotate, and drop blocks.

thermal block: A space or spaces in a building having similar space conditioning requirements so that those areas can be maintained with a single thermal control system.

thermistor: A type of resistor whose resistance is dependent on temperature.

thermostat and occupancy sensor schedule: The thermostat and occupancy sensor schedule indicates each thermostat's unique identifier, room number and control features and functions.

thermostat hacks: Intelligent thermostats can track a user's heat and air-conditioning habits, learn user preferences, and generally surveil a location remotely. One HVAC controls manufacturer has programmed their thermostats to report temperature settings over the Internet to a company database every 12 seconds. In addition, there has been at least one instance where a thermostat produced in the Far East was manufactured so it was retasked remotely (over the Internet) to eavesdrop on sensitive conversations in a conference room.

thermowell: A tubular fitting used to protect temperature sensors installed in industrial processes. It consists of a tube closed at one end and mounted in the process stream. A temperature sensor such as a thermometer, thermocouple or resistance temperature detector is inserted in the open end of the tube, which is usually in the open air outside the process piping or vessel and any thermal insulation. The process fluid transfers heat to the thermowell wall, which in turn transfers heat to the sensor. If the sensor fails, it can be easily replaced without draining the vessel or piping.

threat: Any circumstance or event with the potential to cause harm to a BCS, ICS or SCADA system, adversely impact agency operations (including mission, functions, image, or reputation), agency assets, or individuals through a building control system via unauthorized access, destruction, disclosure, modification of information, and/or denial-of-service. SOURCE: SP 800-53; SP 800-53A; SP 800-27; SP 800-60; SP 800-37; CNSSI-4009

threat agent: An individual, group, organization, or government that conducts or has the intent to conduct cyber-attacks.

threat analysis: The detailed evaluation of the characteristics of individual threats. SOURCE: SP 800-27

threat assessment: Process of formally evaluating the degree of threat to an information system or enterprise and describing the nature of the threat. SOURCE: CNSSI-4009; SP 800-53A

threat model: Describes a given threat and the harm it could to do a BCS if it has a vulnerability.

threat monitoring: Analysis, assessment, and review of audit trails and other information collected for the purpose of searching out building control system events that may constitute violations of building control system security. SOURCE: CNSSI-4009

threat scenario: A set of discrete threat events, associated with a specific threat source or multiple threat sources, partially ordered in time. SOURCE: SP 800-30

threat shifting: Response from adversaries to perceived safeguards and countermeasures (i.e., security controls), in which the adversaries change some characteristic of their intent to do harm in order to avoid or overcome safeguards and countermeasures.

threat source: The intent and method targeted at the intentional exploitation of a vulnerability or a situation and method that may accidentally trigger a vulnerability. Synonymous with *Threat Agent*. SOURCE: FIPS 200; SP 800-53; SP 800-53A; SP 800-37

threat vector: The method a hacker uses to attack the target BCS network.

throughput: Measures the number of service requests on the I/O channel per unit of time.

ticket: In access control, data that authenticates the identity of a client or a service and, together with a temporary encryption key (a session key), forms a credential.

time bomb attack: Resident computer program that triggers an unauthorized act at a predefined time. The "Friday the 13th" computer virus is an example. This virus infects the building control system several days or even months before and lies dormant until the date reaches Friday the 13th. SOURCE: CNSSI-4009

time-dependent password: Password that is valid only at a certain time of day or during a specified interval of time. SOURCE: CNSSI-4009

tiny fragment attack: To impose an unusually small fragment size on outgoing packets. If the fragment size is small enough, a disallowed packet might be passed because it didn't hit a match in the filter.

topology: Geometric arrangement of nodes and cable links in a local area network; may be either centralized or decentralized. Common topologies are a bus, star, and ring.

TP/FT-10: A Free Topology Twisted Pair network defined by CEA-709.3. The most common media type for a CEA-709.1-C control network.

TP/XF-1250: A high speed twisted pair, doubly-terminated bus network defined by the LonMark Interoperability Guidelines. Typically used only to connect multiple TP/FT-10 networks.

traceroute: A network diagnostic tool for displaying the path and measuring transit delays of packets between a computer and a server.

traffic analysis: A form of passive attack in which an intruder observes information about calls (although not necessarily the contents of the messages) and makes inferences; for example, from the source and destination numbers, or frequency and length of the messages. SOURCE: SP 800-24

traffic-flow security (TFS): Techniques to counter Traffic Analysis. SOURCE: CNSSI-4009

traffic light protocol: A set of designations employing four colors (RED, AMBER, GREEN, and WHITE) used to ensure that sensitive information is shared with the properly cleared audience.

traffic padding: Generation of mock communications or data units to disguise the amount of real data units being sent. SOURCE: CNSSI-4009

trampolining: In a buffer overflow attack, if the address of the user-supplied data is unknown, but the location is stored in a register, then the return address can be overwritten with the address of an opcode, which causes execution to jump to the user-supplied data. If the location is stored in a register R, then a jump to the location containing the opcode for a jump R, call R or similar instruction, will cause execution of user-supplied data.

transaction authentication number (TAN): Used by some online banking services as a form of single use one-time passwords to authorize financial transactions. TANs are a second layer of security above and beyond the traditional single-password authentication. TANs provide additional security because they act as a form of two-factor authentication. Should the physical document or token containing the TANs be stolen, it will be of little use without the password; conversely, if the login data are obtained, no transactions can be performed without a valid TAN. Some banks dispatch such TANs to the user's mobile phone via SMS, in which case they are called mTANs (for *mobile TANs*).

transfer rate: The number of megabytes of data that can be transferred from the read/write heads to the disk drive controller in one second.

transformer: An apparatus for reducing or increasing the voltage of an alternating electric current.

Transmission Control Protocol (TCP): TCP enables two hosts to establish a connection and exchange streams of data and ensures data delivery in the correct sequence.

transmission security (TRANSEC): Measures (security controls) applied to transmissions in order to prevent interception, disruption of reception, communications deception, and/or derivation of intelligence by analysis of transmission characteristics such as signal parameters or message externals. TRANSEC is that field of COMSEC that deals with the security of communication transmissions, rather than that of the information being communicated. SOURCE: CNSSI-4009

transmitter: Equipment that generates and transmits a message or signal.

trap door: Bypassing security controls. (1) A means of reading cryptographically protected information by the use of private knowledge of weaknesses in the cryptographic algorithm used to protect the data. (2) In cryptography, one-to-one function that is easy to compute in one direction, yet believed to be difficult to invert without special information. SOURCE: CNSSI-4009

triage: The process of receiving, initial sorting, and prioritizing of information to facilitate its appropriate handling.

triple-wrapped: Data that has been signed with a digital signature, encrypted, and then signed again.

Trojan horse attack: A non-self-replicating program that seems to have a useful purpose, but in reality has a different, malicious purpose. A computer program that appears to have a useful function, but also has a hidden and potentially malicious function that evades security mechanisms, sometimes by exploiting legitimate authorizations of a system entity that invokes the program. SOURCE: SP 800-61; CNSSI-4009

trolling: Internet user behavior meant to intentionally anger or frustrate someone in order to provoke a response.

trunking: Connecting switches together so that they can share VLAN information.

trusted channel: A channel where the endpoints are known and data integrity is protected in transit. Depending on the communications protocol used, data privacy may be protected in transit. Examples include *SSL, IPSEC*, and *secure physical connection*. SOURCE: CNSSI-4009

tunneled password protocol: A protocol where a password is sent through a protected channel. For example, the TLS protocol is often used with a verifier's public key certificate to (1) authenticate the verifier to the claimant, (2) establish an encrypted session between the verifier and claimant, and (3) transmit the claimant's password to the verifier. The encrypted TLS session protects the claimant's password from eavesdroppers. SOURCE: SP 800-63

tunneling: Technology enabling one network to send its data via another network's connections. Tunneling works by encapsulating a network protocol within packets carried by the second network. SOURCE: CNSSI-4009

turnkey: A product or building control system that can be plugged in, turned on and operated requiring little or no additional configuring.

Two-Factor Authentication: Proof of identity by two independent means, such as knowing a password and using a smartcard.

two-part code: Code consisting of an encoding section, in which the vocabulary items (with their associated code groups) are arranged in alphabetical or other systematic order, and a decoding section, in which the code groups (with their associated meanings) are arranged in a separate alphabetical or numeric order. SOURCE: CNSSI-4009

two-person control (TPC): Continuous surveillance and control of positive control material at all times by a minimum of two authorized individuals, each capable of detecting incorrect and unauthorized procedures with respect to the task being performed and each familiar with established security and safety requirements. SOURCE: CNSSI-4009

two-person integrity (TPI): System of storage and handling designed to prohibit individual access by requiring the presence of at least two authorized individuals, each capable of detecting incorrect or unauthorized security procedures with respect to the task being performed. See *no-lone zone*. SOURCE: CNSSI-4009

Type 1 Key: Generated and distributed for use in a cryptographic device for the protection of national security information. SOURCE: CNSSI-4009, as modified

Type 1 Product: Cryptographic equipment, assembly or component for encrypting and decrypting national security information when appropriately keyed. Developed using established business processes and containing approved algorithms. Used to protect systems requiring the most stringent protection mechanisms. SOURCE: CNSSI-4009

Type 2 Key: Generated and distributed for use in a cryptographic device for the protection of unclassified information. SOURCE: CNSSI-4009

Type 2 Product: Cryptographic equipment, assembly, or component for encrypting or decrypting sensitive information when appropriately keyed. Developed using established business processes and containing approved algorithms. Used to protect systems requiring protection mechanisms exceeding best commercial practices including systems used for the protection of unclassified information. SOURCE: CNSSI-4009, as modified

Type 3 Key: Used in a cryptographic device for the protection of unclassified sensitive information, even if used in a Type 1 or Type 2 product. SOURCE: CNSSI-4009

Type 3 Product: Unclassified cryptographic equipment, assembly, or component used, when appropriately keyed, for encrypting or decrypting unclassified sensitive US government or commercial information, and to protect systems requiring protection mechanisms consistent with standard commercial practices. Developed using established commercial standards and containing NIST-approved cryptographic algorithms/modules or successfully evaluated by the National Information Assurance Partnership (NIAP). SOURCE: CNSSI-4009

Type 4 Key: Used by a cryptographic device in support of its Type 4 functionality; that is, any provision of key that lacks US government endorsement or oversight. SOURCE: CNSSI-4009

Type 4 Product: Unevaluated commercial cryptographic equipment, assemblies, or components not certified for government usage. These products are typically delivered as part of commercial offerings and are commensurate with the vendor's commercial practices. These products may contain either vendor proprietary algorithms, algorithms registered by NIST, or algorithms registered by NIST and published in a FIPS. SOURCE: CNSSI-4009

type accreditation: A form of accreditation that is used to authorize multiple instances of a major application or general support system for operation at approved locations with the same type of computing environment. In situations where a major application or general support system is installed at multiple locations, a type accreditation satisfies C&A requirements only if the application or system consists of a common set of tested and approved hardware, software, and firmware. SOURCE: CNSSI-4009

type certification: The certification acceptance of replica information systems based on the comprehensive evaluation of the technical and nontechnical security features of an information system and other safeguards, made as part of and in support of the formal approval process, to establish the extent to which a particular design and implementation meet a specified set of security requirements. SOURCE: CNSSI-4009

CHAPTER 22

U

US person: Federal law and executive order define a US person as a citizen of the United States; an alien lawfully admitted for permanent residence; an unincorporated association with a substantial number of members who are citizens of the United States or are aliens lawfully admitted for permanent residence; and/ or a corporation that is incorporated in the United States. SOURCE: CNSSI-4009

US-controlled facility: Base or building to which access is physically controlled by US individuals who are authorized US government or US government contractor employees. SOURCE: CNSSI-4009

unauthorized access: A person gains logical or physical access without permission to a network, building control system, application, data, or other resource. SOURCE: SP 800-61

unclassified: Information that has not been determined pursuant to E.O. 12958, as amended, or any predecessor order, to require protection against unauthorized disclosure and that is not designated as classified. SOURCE: CNSSI-4009

underfloor air distribution (UFAD): UFAD systems use an underfloor supply plenum located between the structural concrete slab and a raised floor system to supply conditioned air through floor diffusers directly into the occupied zone of the building.

uninterruptible power supply (UPS): A backup power supply with enough power to allow a safe and orderly shutdown of the computer should there be a disruption of commercial power.

universal serial bus (USB): Standard for connecting electronic devices to a computer using a serial bus.

universal software radio peripheral (USRP): A software-defined radio is an inexpensive hardware platform for software radio commonly used by research labs, universities, and hobbyists. A USRP can be used as a transmitter/receiver and decoder; an RFID reader; a GPS; a cellular GSM base station; a digital television (ATSC) decoder; and passive radar.

Unix: An operating system that supports multitasking and is ideally suited to multi-user applications (such as networks). This operating system was originally derived from AT&T Unix developed in 1970 at the Bell Labs research center.

unprotected share: In Windows, a *share* is a mechanism that allows a user to connect to other building control systems. An *unprotected share* is one that allows anyone to connect to it.

unstable: A building control system that is not constant.

untrusted process: Process that has not been evaluated or examined for correctness and adherence to the security policy. It may include incorrect or malicious code that attempts to circumvent the security mechanisms. SOURCE: CNSSI-4009

© Luis Ayala 2016
L. Ayala, *Cybersecurity Lexicon*, DOI 10.1007/978-1-4842-2068-9_22

upgrade: A new release of software, hardware, or firmware to fix errors or vulnerabilities or to provide additional functionality.

uploading: Connecting to another computer and sending a copy of program or file to that computer.

US-CERT: A partnership between the Department of Homeland Security and the public and private sectors, established to protect the nation's Internet infrastructure. US-CERT coordinates defense against and responses to cyber-attacks across the nation. SOURCE: CNSSI-4009

useful records: Records that are helpful but not required on a daily basis for continued building operations.

user: A person or a process accessing an automated building control system, either by direct or indirect connection.

user contingency procedures: Manual procedures to be implemented during a building control system outage.

User Datagram Protocol (UDP): A connection-less transport layer protocol between a network layer and an application layer.

User-defined Configuration Property Type (UCPT): Configuration Property format type defined by the device manufacturer.

User-defined Network Variable Type (UNVT): A network variable format defined by the device manufacturer. UNVTs create non-standard communications (other vendor's devices may not correctly interpret it) and may close the building control system so they are generally not permitted.

user ID: Unique symbol or character string used by an information system to identify a specific user. SOURCE: CNSSI-4009

user initialization: A function in the life cycle of keying material; the process whereby a user initializes its cryptographic application (e.g., installing and initializing software and hardware). SOURCE: SP 800-57

User Partnership Program (UPP): Partnership between US government agencies to facilitate development of secure information system equipment incorporating approved cryptography. The result of this program is the authorization of the product or system to safeguard information in the user's specific application. SOURCE: CNSSI-4009

User-to-Root (U2R) Attack: This cyber-attack occurs when an attacker with access to a normal user account is able to exploit a building control system vulnerability to gain root access.

Utility Control System (UCS): A type of industrial control system. Used for field control of utility systems such as an electrical substation, sanitary sewer lift station, water pump station, and so forth. Building control systems are excluded from a UCS, however it is possible to have a Utility Control System and a Building Control System in the same facility, and for those systems to share components such as the FPOC. A UCS may include its own local front-end.

Utility Monitoring and Control System (UMCS) Network: A network connecting multiple building control networks (BCNs) using the CEA-852-B standard.

CHAPTER 23

V

%VOL: Concentration of gas, measured in percentage by volume.

validate: Process of checking documents or testing against a formal standard.

validation: The process of demonstrating that the system under consideration meets in all respects the specification of that system. Confirmation (through the provision of strong, sound, objective evidence) that requirements for a specific intended use or application have been fulfilled (e.g., a trustworthy credential has been presented, or data or information has been formatted in accordance with a defined set of rules, or a specific process has demonstrated that an entity under consideration meets, in all respects, its defined attributes or requirements). SOURCE: FIPS 201; CNSSI-4009

variable air volume terminal (VAV): Often called a *VAV box* or a *fan-powered terminal unit*, is a zone-level flow control device. It is a calibrated air damper with an automatic actuator. Unlike constant air volume HVAC systems, which supply a constant airflow at a variable temperature, VAV systems vary the airflow at a constant temperature. The advantages of VAV systems over constant air volume systems include more precise temperature control, reduced compressor wear, lower energy consumption, less noise and passive dehumidification.

value-added reseller (VAR): A business that repackages and improves hardware manufactured by an original equipment manufacturer.

valve: An in-line device in a fluid-flow system that can interrupt flow, regulate the rate of flow, or divert flow to another branch of the system. There are many different types and styles of valves, but all primarily serve the common purpose of balancing a system. Valve types include the following:

- **Three-way valves**. Most associated with constant volume systems, these devices are used to modulate water flow to the load without changing the constant volume of water flow to the system.

- **Two-way valves**. Most associated with variable speed/variable volume systems, these devices modulate flow to the load by changing the constant volume of water flow to the system.

- **Manual balancing valves**. These have an adjustable orifice that can be changed by hand to provide a specific pressure drop and flow.

- **Flow-limiting valves**. These valves vary the flow based on differential pressure to provide a specific flow rate.

SOURCE: NIST SP 800-82

© Luis Ayala 2016
L. Ayala, *Cybersecurity Lexicon*, DOI 10.1007/978-1-4842-2068-9_23

valve schedule: The valve schedule indicates each valve's size, flow coefficient Kv (Cv), pressure drop at specified flow rate, spring range, positive positioner range, actuator size, close-off pressure to torque data, dimensions, and access and clearance requirements data. The valve schedule includes actuator data of the force required to move and seal the valve.

valve settings: Closing valves at certain points in time could lead to an increase of pressure that could damage equipment. This was one technique used by the Stuxnet virus.

vampire tap: A device for physically connecting a station (e.g., a computer or printer) to a network that uses 10BASE5 cabling. This device clamps onto and "bites" into the cable (hence the vampire name), forcing a spike through a hole drilled through the outer shielding to contact the inner conductor while other spikes bite into the outer conductor. Vampire taps allow new connections to be made on a given physical cable while the cable is in use. Also called a *piercing tap*.

vaporware: Marketing materials describing computer technology that does not exist (in fact).

variable frequency drive (VFD): (1) A way to control the rotational speed of an alternating current (AC) electric motor by controlling the frequency of the electrical power supplied to the motor. (2) A type of motor controller that drives an electric motor by varying the frequency and voltage supplied to the electric motor. Other names for a VFD are *variable speed drive, adjustable speed drive, adjustable frequency drive, AC drive, microdrive*, and *inverter*.

Vehicular Ad hoc Networks (VANETs): Used for communication between vehicles and roadside equipment.

verification: Confirmation, through the provision of objective evidence, that specified requirements have been fulfilled (e.g., an entity's requirements have been correctly defined, or an entity's attributes have been correctly presented; or a procedure or function performs as intended and leads to the expected outcome). SOURCE: CNSSI-4009

verifier impersonation attack: An attack where the attacker impersonates the verifier in an authentication protocol, usually to learn a password. SOURCE: SP 800-63

vibrational energy harvester (VEH): A small piezoelectric device using micro-electro-mechanical systems technology that is capable of harvesting microwatts of electricity from background vibrations to power miniature devices like sensor nodes. For example, a VEH was attached to a wireless temperature sensor and subjected to vibrations of 353 Hz at 0.64g. The system generated enough power to take environmental readings and send data to a base station at 15-second intervals.

Virtual Private Network (VPN): A virtual network—built on top of existing physical networks—that provides a secure communications tunnel for data and other information transmitted between networks. Protected information system link utilizing tunneling, security controls (see *information assurance*), and endpoint address translation giving the impression of a dedicated line. SOURCE: SP 800-46; CNSSI-4009

virtuous circle and vicious circle: A complex chains of events that reinforce themselves through a feedback loop. A virtuous circle has favorable results, while a vicious circle has detrimental results.

virus attack: Software buried within an existing program designed to infect a computer. A code segment that replicates by attaching copies of itself to existing executable programs. This is usually done in such a manner that the copies will be executed when the file is loaded into memory, allowing them to infect still other files. The new copy of the virus is executed when a user executes the new host program. The virus may include any additional "payload" that is triggered when specific conditions are met. For example, some viruses display a text string on a particular date. There are many types of viruses including variants, overwriting, resident, stealth, and polymorphic. Viruses often have damaging side effects, sometimes intentionally, sometimes not. SOURCE: NIST Bulletin

virus definitions: Predefined signatures for known malware used by antivirus detection algorithms.

virus detection software: Software written to scan machine-readable media on building control systems. There are a growing number of reputable software packages available that are designed to detect or remove viruses. In addition, many utility programs can search text files for virus signatures or potentially unsafe practices.

virus hoax: An urgent warning message about a nonexistent virus. SOURCE: SP 800-61

virus signature: A unique set of characters, which identify a particular virus. This may also be referred to as a *virus marker.*

vishing attack: A type of phishing attack that uses a telephone ("v" is for voice) to obtain personal information. Phishing target is called directly by criminals or receives an e-mail asking the target to call a specific phone number.

visual malware: A novel Trojan horse that allows a hacker to engage in remote reconnaissance through use of a smartphone's camera and other sensors to obtain geolocation data and its accelerometer to create a 3D map of the phone's surroundings. Using PlaceRaider for example, a hacker can download images of the physical space, study the environment and carefully construct a three dimensional model of indoor environments to surveil the target's private home or work space. PlaceRaider can be used to steal virtual objects from the environment such as financial documents, information on computer monitors, and personally identifiable information.

visual microphones hack: MIT researchers claim to have created an algorithm that can reconstruct sound (and even intelligible speech) with the tiny vibrations it causes on video (as small as thousandths of a pixel). When signals were averaged, researchers were able to extract sound that makes sense. In the example compilation video, a bag of chips is filmed from 15 feet away, through sound-proof glass. Although in the reconstructed audio the words being said are not clear, the words are possible to decipher. A high-speed camera (2,000: 6,000 frames/sec) was used, however cheap cameras could be used by taking advantage of a bug called *rolling shutter* that encodes information at a much higher rate than the camera's actual frame rate. This allows recovery of sound at frequencies several times higher than the frame rate of the camera.

vital records: Records or documents, for legal, regulatory, or operational reasons, cannot be lost or damaged without materially impairing the organization's ability to operate.

voice intrusion prevention system (VIPS): A security management system for voice networks that monitors voice traffic for multiple calling patterns or cyber-attack signatures to detect anomalous behavior.

Voice over IP (VoIP): The ability to carry voice traffic alongside data traffic on a LAN or the Internet.

voice recovery: The restoration of an organization's voice communications system.

voltage reference: An electronic device that ideally produces a fixed (constant) voltage irrespective of the loading on the device, power supply variations, temperature changes, and the passage of time. Voltage references are used in power supplies, analog-to-digital converters, digital-to-analog converters, and other measurement and control systems.

vulnerability: A weakness in a building control system or component (e.g., security procedures, hardware design, and internal controls) that could be exploited or attacked, or fail. Vulnerabilities include susceptibility to physical dangers, such as fire or water, unauthorized access to sensitive data, entry of erroneous data, denial of timely service, fraud, and so forth. SOURCE: SP 800-53; SP 800-53A; SP 800-37; SP 800-60; SP 800-115; FIPS 200

vulnerability assessment: Formal description and evaluation of the vulnerabilities in an information system. Systematic examination of an information system or product to determine the adequacy of security measures, identify security deficiencies, provide data from which to predict the effectiveness of proposed security measures, and confirm the adequacy of such measures after implementation. SOURCE: SP 800-37; SP 800-53; SP 800-53A; CNSSI-4009 Three categories of vulnerability assessments are possible based on how much information is provided to the test team performing the assessment.

1. **White-Box Test Team**: Test team has complete and unrestricted access on site to the BCS network including network diagrams, hardware, operating system and application details. Knowledge of network allows the team to target specific building equipment, applications and field devices.

2. **Grey-Box Test Team**: Test team simulates attack by a disgruntled employee on site. Test team has user-level privileges and access permitted to the BCS network with certain security policies relaxed.

3. **Black-Box Test Team**: Test team has no prior knowledge of BCS network (except possibly a web site URL or IP address). Test team attempts to break into the BCS network remotely.

vulnerability assessment & management: In cybersecurity work, a person that conducts assessments of threats and vulnerabilities, determines deviations from acceptable configurations, enterprise or local policy, assesses the level of risk, and develops and/or recommends appropriate mitigation countermeasures in operational and non-operational situations. SOURCE: NICCS

vulnerability stockpiling: The concept that rather than disclosing vulnerabilities to software makers so that they can be patched, a government might buy and then stockpile zero-days for possible use as an offensive weapon.

vulnerability testing: The intent of vulnerability testing is to identify potential problems with a building controls system. There are at least five types of vulnerability tests. Tests are conducted monthly, quarterly, or annually.

- **external vulnerability scan**: Identify network-facing vulnerabilities (monthly).

- **internal vulnerability scan**: Identify network-facing vulnerabilities (quarterly).

- **external vulnerability assessment**: Identify configuration and architecture vulnerabilities (annual).

- **internal vulnerability assessment**: Identify network, client, configuration and physical vulnerabilities (annual).

- **penetration test**: Exploit any vulnerability to obtain access to building controls (annual).

CHAPTER 24

W

wall wart: A small external power supply in a plastic case that plugs into a wall outlet. Other common names include *plug pack, plug-in adapter, AC adapter, AC/DC adapter, AC/DC converter, adapter block, line power adapter, power brick* and *power adapter*.

war chalking: Marking areas on sidewalks with chalk to indicate where wireless networks can be accessed.

war dialer: A computer program that dials a series of telephone numbers to find lines connected to building control systems, and catalogs those numbers so that a hacker can break in.

war dialing attack: Dialing all the telephone numbers in a given area code to locate BCS devices connected by a modem.

war droning attack: Use of a cyber-drone to search for Wi-Fi wireless networks connected to a BCS at a facility and hack into BCS networks when they are found. A cyber-drone can also shut down computer systems and other nearby electronic systems from the sky through targeted emission of microwaves. *See Counter-electronics High-powered Microwave Advanced Missile Project (CHAMP)*.

Wardriving Attack: The act of searching for Wi-Fi wireless networks by a hacker in a moving vehicle, using a portable computer, smartphone or personal digital assistant (PDA). Also called *access point mapping*.

WARNORD: Warning Order issued by the United States Cyber Command in response to a suspected cyber-attack.

war walking attack: Just like wardriving (searching for Wi-Fi wireless networks), but on foot.

warm site: An environmentally conditioned workspace that is partially equipped with information systems and telecommunications equipment to support relocated operations in the event of a significant disruption. Backup site that typically contains the data links and preconfigured equipment necessary to rapidly start operations, but does not contain live data. Thus commencing operations at a warm site requires (at a minimum) the restoration of current data. SOURCE: SP 800-34; CNSSI-4009

watchdog timer: A watchdog timer (sometimes called a *computer operating properly* or *COP timer*) is an electronic timer that is used to detect and recover from building control system malfunctions. During normal operation, the building control system regularly restarts the watchdog timer to prevent it from elapsing, or "timing out." If, due to a hardware fault or program error, the building control system fails to restart the watchdog, the timer will elapse and generate a timeout signal.

water holing attack: A hacker will set up a fake web site or compromise a legitimate one in order to exploit visiting users.

waterside economizer: Provides chilled water to a building using one or more cooling towers and heat exchangers. A waterside economizer sequencer module is designed to automatically switch the central plant chiller (mechanical) cooling and waterside economizer with minimal disturbance to flows and to the chilled water supply temperature.

warez d00dz: A cracker subculture that gets illegal copies of copyrighted software. If it has copy protection on it, they break the protection so the software can be copied.

web bug: A tiny image (malicious code), invisible to a user, placed on web pages in such a way to enable third parties to track use of web servers and collect information about the user, including IP address, host name, browser type and version, operating system, building control system, and cookies. SOURCE: SP 800-28

web cache: A cache of fill requests from a web server that are stored locally to reduce Internet traffic.

web content filtering software: A program that prevents access to undesirable web sites, typically by comparing a requested web site address to a list of known bad web sites. SOURCE: SP 800-69

webcam hack: Most webcams can be hacked. A hacker can watch your facility without your knowledge. This is a fairly simple hack made possible by a Trojan horse called Blackshades that even a script kiddie can master. What's worse is a hacker may be able to hack into your BCS through IP-enabled cameras.

webcrawler: A program that automatically searches the Web to find new pages to add to search engines. Also called a *spider*.

web log sensor: Reports information from Apache, IIS and SSL error logs. Provides information about suspicious URLs as well as web server errors.

web site: A location on the World Wide Web that is owned and managed by an individual, company or organization; usually contains a home page and pages that include information provided by the site's owner, and may include links to other relevant sites.

website reputation service: Maintains a blacklist of known bad web sites.

WeMo: Innovative, easy-to-use products developed by Belkin that use mobile Internet to control home electronics, power, water and Wi-Fi right from a smartphone or tablet. WeMo also works with IFTTT, connecting home electronics to a whole world of online apps. One example is the WeMo Insight Switch that sends notifications to a smartphone. It has the ability to turn home electronics on and off, monitor their behavior and energy consumption.

wet stacking: A condition in diesel engines in which not all the fuel is burned and passes on into the exhaust side of the turbocharger and on into the exhaust system. A diesel engine drips a thick, dark substance from its exhaust pipes. Typically caused by operating the engine at light load for an extended period. This is a progressive condition that will lead to more wet stacking and prolonged operation at low loads can lead to permanent engine damage.

wetware: A term used to describe the elements equivalent to hardware and software found in a person, namely the central nervous system and the human mind. The prefix *wet* is a reference to the water found in living creatures.

whaling attack: Spear phishing targeting high-profile executives, politicians, and celebrities. Whaling e-mails are highly-personalized and appear to come from a trusted source. Once opened, the target is directed to a web site that was created specifically for that individual's attack. Successful whaling targets are referred to as having been *harpooned*.

White Team: (1) The group responsible for refereeing an engagement between a Red Team of mock attackers and a Blue Team of actual defenders of their enterprise's use of information systems. In an exercise, the White Team acts as the judges, enforces the rules of the exercise, observes the exercise, scores teams, resolves any problems that may arise, handles all requests for information or questions, and ensures that the competition runs fairly and does not cause operational problems for the defender's mission. The White Team helps to establish the rules of engagement, the metrics for assessing results and the procedures for providing operational security for the engagement. The White Team normally has responsibility for deriving lessons-learned, conducting the post engagement assessment, and promulgating results. (2) Can also refer to a small group of people who have prior knowledge of unannounced Red Team activities. The White Team acts as observers during the Red Team activity and ensures the scope of testing does not exceed a predefined threshold. SOURCE: CNSSI-4009

white list: a list or register of entities that are being provided a particular privilege, service, mobility, access or recognition. Entities on the list will be accepted, approved and/or recognized. Whitelisting is the reverse of blacklisting, the practice of identifying entities that are denied, unrecognized, or ostracized. This helps to stop the execution of malware, unlicensed software, and other unauthorized software.

wide area network (WAN): A network that uses high-speed, long-distance communications technology (e.g., phone lines and satellites) to connect computers over long distances. Similar to a LAN, except that parts of a WAN are geographically separated, possibly in different cities or even on different continents. Telecommunications carriers are included in most WANs; very large WANs incorporate satellite stations or microwave towers.

Wi-Fi Direct: A Wi-Fi standard enabling devices to easily connect with each other without requiring a wireless access point.

Wi-Fi hotspot: A physical location that offers Internet access over a wireless local area network (WLAN) through the use of a router connected to a link to an Internet Service Provider. Long range antennas can be hooked up to laptop computers with an external antenna jack: these allow a user to pick up a signal from as far as several kilometers away.

Wi-Fi Protected Access (WPA2): The approved Wi-Fi Alliance interoperable implementation of the IEEE 802.11i security standard. For federal government use, the implementation must use FIPS-approved encryption, such as AES. SOURCE: CNSSI-4009

Wireless Display (WiDi): Enables users to stream music, movies, photos, videos and apps wirelessly from a compatible computer to a compatible HDTV or through the use of an adapter with other HDTVs. WiDi supports HD 1080p video quality, 5.1 surround sound, and low latency for interacting with applications that are sent to the TV from a PC.

Wireless Gigabit (WiGig): WiGig allows devices to communicate high performance wireless data, display and audio without wires at multi-gigabit speeds. WiGig tri-band enabled devices operate in the 2.4, 5 and 60 GHz bands and deliver data transfer rates up to 7 Gbit/s, while maintaining compatibility with existing Wi-Fi devices. The 60 GHz signal cannot typically penetrate walls but can propagate off reflections from walls, ceilings, floors and objects using beamforming built into the WiGig system. When roaming away from the main room the protocol can switch to make use of the other lower bands at a much lower rate, but which can propagate through walls.

WirelessHD: A proprietary standard for wireless transmission of high-definition video content for consumer electronics products at up to 10 meters. The atmospheric absorption of 60 GHz energy by oxygen molecules limits undesired propagation over long distances and helps control intersystem interference and long distance reception, which is a concern to video copyright owners. Also known as *UltraGig*.

Wiper virus: Cyber-attack software that deletes the content of every hard drive it touches. Typically associated with hacktivism; that is, deletion of data and backups for strategic political or wartime goals.

Wired Equivalent Privacy (WEP): A security protocol, specified in the IEEE 802.11 standard, that is designed to provide a WLAN with a level of security and privacy comparable to what is usually expected of a wired LAN. WEP is no longer considered a viable encryption mechanism due to known weaknesses. SOURCE: SP 800-48

Wireless Access Point (WAP): A device that acts as a conduit to connect wireless communication devices together to allow them to communicate and create a wireless network. SOURCE: CNSSI-4009

wireless ad hoc network (WANET): A decentralized type of wireless network. The network is ad hoc because it does not rely on a pre-existing infrastructure, such as routers in wired networks or access points in managed (infrastructure) wireless networks. Instead, each node participates in routing by forwarding data for other nodes, so the determination of which nodes forward data is made dynamically on the basis of network connectivity. Microsoft does not allow advanced encryption and security protocols for wireless ad hoc networks on Windows. In fact, the security hole provided by Ad hoc networking is not only the ad hoc network itself, but the bridge it provides into other networks. In addition to the classic routing, ad hoc networks can use flooding for forwarding data. There are several types of wireless ad hoc networks:

- Mobile Ad hoc Networks (MANET)
- Vehicular Ad hoc Networks (VANETs)
- SmartPhone Ad hoc Networks (SPANs)
- Internet-based Mobile Ad hoc Networks (iMANETs)
- Military / Tactical MANETs

wireless device: A device installed on a BCS that can connect to a manufacturer's database system via radio or infrared waves to collect data, but also to modify control set points. *Something to be removed or disabled immediately if you have any sense at all.*

Wireless Fidelity (Wi-Fi): Any type of 802.11 network.

wireless interface module: Provides a common interface to wireless sensors and switches. These should be extremely rare in a building control system because they are not secure and easily compromised by any garden-variety hacker, junior-grade.

Wireless Sensor Network (WSN) cyber-attack: These cyber-attacks prevent sensors from detecting and transmitting data through the building automation network infrastructure.

wireless signature: Every wireless device has a unique hardware signature assigned to it by the manufacturer. These signatures are broadcast by wireless devices as they probe for, connect to, and use wireless networks.

Wireshark: A network protocol analyzer for providing details about your network protocols, packet information, decryption, and so forth.

Wisconsin Protocol: Instructions for cleaning air conditioning condensate trays that calls for an initial shock treatment with 50 ppm free residual (total) chlorine, addition of detergent to disperse bio-fouling, maintenance of 10 ppm chlorine for 24 hours, and a repeat of the cycle until there is no visual evidence of biofilms. To prevent exposure during cleaning and maintenance, technicians wear proper personal protective equipment: a Tyvek-type suit with a hood, protective gloves, and a properly fitted respirator with a high-efficiency particulate (HEPA) filter or a filter effective at removing one-micron particles.

WITCHCOVEN: Malware that uses profiling techniques to collect technical information on a user's computer to smartly tailor targeted operations. Cyber threat actors are building profiles of potential victims and learning about the vulnerabilities in users' computers to identify victims and tailor future infection attempts. When an unsuspecting user visits a compromised web site, a small piece of inserted code—embedded in the site's HTML and invisible to casual visitors—quietly redirects the user's browser to a second compromised web site without the user's knowledge. The second web site hosts the WITCHCOVEN script, which collects technical information on the user's computer. As of early November 2015, FireEye identified a total of 14 web sites hosting the WITCHCOVEN profiling script.

WokFi: A slang term for a style of homemade Wi-Fi antenna consisting of a crude parabolic antenna made with a low-cost Asian kitchen wok, spider skimmer, or similar household metallic dish. The dish forms a directional antenna that is pointed at the wireless access point antenna, allowing reception of the wireless signal at greater distances than standard omnidirectional Wi-Fi antennas.

work factor: An estimate of the effort or time needed by a potential adversary, with specified expertise and resources, to overcome a protective measure. SOURCE: CNSSI-4009

workstation: A computer used for tasks such as programming, engineering, and design.

World Wide Web (WWW): A global hypertext system operating on the Internet that enables electronic communication of text, graphics, audio, and video.

worm attack: A self-replicating, self-propagating, self-contained program that uses networking mechanisms to spread itself that harms the network and consumes bandwidth. SOURCE: SP 800-61; CNSSI-4009

wrapper: A type of malware concealed inside a legitimate software program to make it undetectable.

write-blocker: A device that allows investigators to examine media while preventing data writes from occurring on the subject media. SOURCE: SP 800-72

CHAPTER 25

X, Y

X.509 Certificate: The International Organization for Standardization/International Telecommunication Union: Standardization Department (ISO/ITUT) X.509 standard defined two types of certificates: the X.509 public key certificate and the X.509 attribute certificate. Most commonly, an X.509 certificate refers to the X.509 public key certificate. SOURCE: SP 800-57

Xcode: Apple development tool.

XcodeGhost: A counterfeit version of an Apple development tool, *Xcode*, downloaded by developers from third-party sources, because downloading the 4GB code from Apple took too long. XcodeGhost poses a privacy risk, as apps developed with XcodeGhost could be configured to record data from people's devices and sent to a remote server.

XOR engine: Process or set of instructions that calculates data bit relationships in a RAID subsystem.

CHAPTER 26

Z

ZeroAccess attack: A Trojan horse bot used to download other malware on an infected machine from a botnet, while remaining hidden on a building control system using rootkit techniques.

zero-day exploit attack: A worm, virus, or other cyber-threat that hits users on the same day the vulnerability is announced.

zero fill: To fill unused storage locations in a building control system with the representation of the character denoting 0. SOURCE: CNSSI-4009

zeroize: To remove or eliminate the key from a cryptographic equipment or fill device. SOURCE: CNSSI-4009

zeroization: A method of erasing electronically stored data, cryptographic keys, and CSPs by altering or deleting the contents of the data storage to prevent recovery of the data. SOURCE: CNSSI-4009

zero net energy (ZNE): A *zero-energy building*, also known as a *zero net energy* (ZNE) building, *net-zero energy building* (NZEB), or *net zero building* is a building with zero net energy consumption, meaning the total amount of energy used by the building on an annual basis is roughly equal to the amount of renewable energy created on the site.

Zipper Effect: Certain load-bearing structures with discrete structural components can be subject to a failure when a single structural member increases the load on adjacent members. In the case of the Hyatt Regency walkway collapse, a suspended walkway (which was already overstressed due to an error in construction) failed when a single vertical suspension rod failed, overloading the neighboring rods which failed sequentially (i.e., like a zipper). Properly designed structures use an adequate factor of safety and/or alternate load paths to prevent this type of mechanical cascade failure.

zombie attack: Synonym: *bot*.

zone: A distinct physical area in which alarm, supervisory, monitor, security, signal, paging, telephone or relay devices are located.

zone disconnect: A way to disconnect a device, circuit, zone, or operation from the building control system programming so that any activations that would normally affect the building control system are ignored and related actions are not executed. Used to disable a circuit or device so that the remainder of the building control system can continue to operate normally.

zone of control: Three-dimensional space surrounding equipment that processes classified and/or sensitive information within which TEMPEST exploitation is not considered practical or where legal authority to identify and remove a potential TEMPEST exploitation exists. SOURCE: CNSSI-4009

© Luis Ayala 2016
L. Ayala, *Cybersecurity Lexicon*, DOI 10.1007/978-1-4842-2068-9_26

CHAPTER 27

■ ■ ■

Facilities, Engineering, and Cyber Acronyms

AAL	Advanced Analytical Laboratory
ABC	above ceiling
AC	alternating current
A/C	air conditioning
ACI TTP	Advanced Cyber Industrial Control System Tactics, Techniques, and Procedures
ACL	access control list
AES	Advanced Encryption Standard
AFCI	arc fault circuit interrupter
AFF	above finished door
AFG	above finished grade
AGA	American Gas Association
AGC	application generic controller
AHJ	authority having jurisdiction
AHU	air handling unit
AIS	alarm indication station
AL	aluminum
ALC	area lighting controller
AMAN	arrival manager
AMB	ambient
AMI	advanced metering Infrastructure
AMP	ampere
ANT	analyst network tool
AO	authorizing official
AOR	area of responsibility
API	American Petroleum Institute
APT	advanced persistent threats
ARP	address resolution protocol
ARR	arrangement
ASA	adaptive security appliance
ASC	application specific controllers
ASTM	American Society for Testing and Materials
ASHRAE	American Society of Heating, Refrigerating and Air Conditioning Engineers
AS&W	attack sensing and warning
ATC	automatic temperature control, at ceiling, Air Traffic Control
AT/FP	Anti-Terrorism/Force Protection

© Luis Ayala 2016
L. Ayala, *Cybersecurity Lexicon*, DOI 10.1007/978-1-4842-2068-9_27

ATM	atmosphere, asynchronous transfer mode
AUTO	automatic
AUX	auxiliary
AVG	average
AV	antivirus
A/V	audio visual
AWS	Autonomous Weapons Systems
B2B	business to business
BAS	Building Automation System
BCP	Business Continuity Planning
BCN	Building Controls Network
BCS	building control system
BBD	boiler blowdown
BDA	battlefield damage assessment
BDD	back draft damper
BF	boiler feed
BFL	bird fancier's lung
BFT	Byzantine fault tolerance
BGP	Border Gateway Protocol
BHP	boiler horsepower
BIA	building impact analysis
BIBBs	BACnet Interoperability Building Blocks
BIM	Building Information Modeling
BLOB	Binary Large OBject
BIoT	Building Internet of Things
BMS	building management system
BOC	building operations center
BOD	bottom of duct
BoE	body of evidence
BOP	bottom of pipe
BOT	bottom
BP	back pressure
B/P/C/S	base/post/camp/station
BPOC	building point of connection
BRP	building resumption planning
B & S	bell and spigot
BSMT	basement
BTU	British thermal unit
BUR	built up roofing
BV	butterfly valve
C2	command and control
C3	command, control and communication
C	condensate line
C to C	center to center
CA	compressed air
CAC	Call Admission Control, Common Access Card, Command and/or Control Center
CAD	computer-aided design
CAFM	computer-aided facility management
CAGE	commercial and government entity code
CAL	calorie
CANBUS	controller area network

CAP	capacity
CAT	category
CAV	constant air volume
CC	coordinating center
CCC	Chaos Computer Club (German)
CCNA	Cisco Certified Network Associate
CD	condensate drain
CDC	Cleared Defense Contractor, Centers for Disease Control
CDRP	Certified Disaster Recovery Planner
CDS	cross-domain solution
CEH	Certified Ethical Hacker
CERT	computer emergency readiness team
CEIR	Correlating Extensive Incident Response
CF	chemical feed, cubic foot
CFH	cubic feet per hour
CFM	cubic feet per minute
CGI	Common Gateway Interface
ChemITC	Chemical Information Technology Council
CHW	chilled water
CI	cast iron, counterintelligence, critical infrastructure
CIA	confidentiality, integrity, availability
CIDX	Chemical Industry Data Exchange
CIO	chief information officer
CIP	Common Industrial Protocol, Critical Infrastructure Protection
CIPAC	Critical Infrastructure Partnership Advisory Council
CIRC	circular, Cyber Incident Response Capability
CIRT	computer incident response team
CISSP	Certified Information Systems Security Professional
CJCSM	Chairman of Joint Chiefs of Staff Manual
CL	center line
CLEC	Competitive Local Exchange Carrier
CM	centimeter, configuration management
CM^2	square centimeter
CMMS	Computerized Maintenance Management System
CMP	central mechanical plant
CMSS	Common Misuse Scoring System
CNA	computer network attack
CNC	computer numerical control
CND	Computer Network Defense
CNDRA	Computer Network Defense Response Actions
CNDSP	Computer Network Defense Service Provider
CNE	Computer Network Exploitation
CNO	Computer Network Operations
CNSSI	Committee on National Security Systems Instruction
CO	clean out
CO^2	carbon dioxide
COA	course of action
col	column
CONC	concrete, concentric
CONN	connect, connection
CONT	continuation

COOP	continuity of operations
COP	common operating picture
C-OPE	Cyber Operational Preparation of the Environment
COTS	commercial off-the-shelf
CP	certificate policy
CPNI	Centre for the Protection of National Infrastructure
CPT	cyber protection team
CPVC	chlorinated polyvinyl chloride
CR	condenser return
CRAC	computer room air conditioner
CRC	cyclic redundancy check
CRL	certificate revocation list
CRR	Cyber Resilience Review
CRT	computer recovery team
CRW	chemical resistant waste
CS	condenser supply
CSA	Cognizant Security Agency
CSET	Cyber Security Evaluation Tool
CSIH	Computer Security Incident Handler
CSIRT	Computer Security Incident Response Team
CSN	central services node
CSO	Cognizant Security Office
CSP	Commerce Service Provider
CSRF	cross-site request forgery
CSSP	Control System Security Program
CT	cooling tower
CTI	Computer Telephony Integration
CTIIC	Cyber Threat Intelligence Integration Center
CTM	City Tie Module
CTO	communications tasking order
CTR	commercial telecom room
CTR	center
CU	cubic, condensing unit
CU FT.	cubic feet
CU IN.	cubic inches
CUI	controlled unclassified information
CUP	central utility plant
CV	check valve
CVE	Common Vulnerabilities and Exposure
CVR	cloud video recording
CVSS	Common Vulnerability Scoring System
CW	condenser water
CW	cold water
CWS	cooling water supply
CWR	cold water riser, cooling water return
CYBERCON	cyber condition
D	deep
DAA	Designated Accrediting Authority (the Authorizing Official)
DAR	data at rest
DAT	digital audio tape
DB	dry bulb, decibels

DC3	Defense Cyber Crime Center
DCO	Dial Entry Office
DCP	Defense Community Planning
DCS	distributed control function
DCV	demand control ventilation
ddc	defensive counter-cyber, direct digital control
DDS	Digital Data Storage
DDoS	distributed denial-of-service attack
DEG	degree
°C	degrees Celsius
°F	degrees Fahrenheit
DELTAT	temperature difference
DER	distributed energy resources
DET	detail
DFS	detailed feasibility study
dg	distributed Generation
DGM	distributed grid management
DGR	dynamic growth and reconfiguration
DHS	Department of Homeland Security
DIA	Defense Intelligence Agency
dia	diameter
DISC	disconnect
DLL	dynamic-link library
DLT	digital linear tape
DMAN	departure manager
DMZ	demilitarized zone
DN	down
DNP3	distributed network protocol
DOAS	dedicated outdoor air system
DOD	Department of Defense
DODI	Department of Defense Instruction
DoS	denial-of-service attack
DP	dew point temperature
DPA	differential power analysis
DPCD	digital protection and control devices
DPI	deep packet inspection
DR	drain, demand reduction
DRAS	demand reduction automation Server
DRP	disaster recovery plan
DSA	Digital Signature Algorithm
DSN	Defense Switch Network
DSS	Defense Security Service
DTG	date-time group
DTN	delay-tolerant networking
DWG	drawing
EA	exhaust air, Enterprise Architecture
EAP	Extensible Authentication Protocol
EAT	entering air temperature
E to C	end to center
ECC	energy consumption control
ECM	energy conservation measures

ECPA	Electronic Communications Privacy Act
EER	energy efficient ratio
EFF	efficiency
EGP	Exterior Gateway Protocol
EIA	Electronic Industries Association
EJ	expansion joint
EL	elevation
ELB	elbow
ELEC	electrical
EMCS	energy management controls system
EMI	electromagnetic interference
EMS	energy management system
ENISA	European Network and Information Security Agency
ENT	entering
EPDM	ethylene propylene diene monomer (rubber)
EPMS	electrical power monitoring system
ERP	enterprise resource planning
ERTMS	European Rail Traffic Management System
ERU	energy recovery unit
ESG	enterprise sensor grid
ESN	Enterprise Storage Network
ESS	electronic security system
ESP	external static pressure
ET	expansion tank
EUI	Energy Use Index
EVAP	evaporator
EWT	entering water temperature
EXH	exhaust
EXP	expansion
EXST	existing
EXT	external
°F	degrees Fahrenheit
F	Fahrenheit
FAR	false acceptance rate
FC	flexible connector, flexible connection, Fiber Channel
FC-AL	fiber channel-arbitrated loop
FCC	Fiber Channel Community
FCL	facility (security) clearance
FCO	floor cleanout
FCS	field control system
FCU	fan coil unit
FDDI	fiber distributed data interface
FD	floor drain
FDE	full disk encryption
FDW	feed water
FEC	fire extinguisher cabinet
FF	finish floor
FG	finish grade
FHC	fire hose cabinet
FIM	file integrity monitoring
FIPS	Federal Information Processing Standard

FIRST	Forum of Incident Response and Security Teams
FISMA	Federal Information Security Management Act
FLA	full load amps
FLR	floor
FM	flow meter
FMC	fully mission-capable
FO	fuel oil
FOV	flush out valve
FPCON	force protection condition
FPM	feet per minute
FPS	feet per second
FRAGORD	fragmentary order
FRR	false rejection rate
FS	flow switch
FSO	facility security officer
FSS	Foreign Security Service
FSSS	frequency hopping spread spectrum
FSW	friction stir welding
FT	foot, feet
ft²	square feet
FTG	fitting
FU	fixture unit
FV	flush valve
G	gram, gas line
GA	gauge
GAL	gallons
GPPC	general purpose programmable controller
GALV	galvanized
GF	ground fault
GFIRST	Global Forum of Incident Response and Security Teams
GICSP	Global Industrial Cyber Security Professional
GIG	global information grid
GII	global information infrastructure
GHDB	Google Hacking Database
GL.V	globe valve
GNCC	Global Network Operations Control Center
GND	ground
GPD	gallons per day
GPH	gallons per hour
GPM	gallons per minute
GPS	gallons per second
GR	grain
gsf	gross square foot
GSS	General Support Systems
GV	gate valve
GWH	gas water heater
HAG	high assurance guard
HBA	host bus adapter
HQ	headquarters
HSPD	Homeland Security Presidential Directive
HVAC	heating, ventilation and air conditioning

H_2O	water
HB	hose bib
HD	head
Hg	mercury
HGT	height
HIDS	host-based intrusion detection system
HMD	humidity
HMI	human-machine interface
H-O-A	Hand-Off-Auto switch
HORIZ	horizontal
HP	horsepower
HR	hour
HSM	hierarchical storage management
HTD	heated
HTR	heater
HVAC	heating ventilation and air conditioning
HW	hot water
HWH	hot water heater
HWR	hot water return
HWS	hot water supply
HWT	hot water tank
Hz	hertz
I&W	indications and warning
IA	information assurance
IAC	information assurance component
IAM	information assurance manager
IAPC	Information Assurance Protection Center
IATT	interim authorization to test
IAVA	information assurance vulnerability alert
IAVM	information assurance vulnerability management
IAW	in accordance with
IC	intelligence community
ICCB	insulated case circuit breaker
ICCP	Inter-Control Center Communications Protocol
ICCWG	nternational CND Coordination Working Group
IC-IRC	Intelligence Community–Incident Response Center
ICMP	Internet Control Message Protocol
ICS	industrial control systems
ICS-CERT	Industrial Control Systems Cyber Emergency Response Team
ICSJWG	Industrial Control Systems Joint Working Group
IDS	intrusion detection system
ID	inside diameter
IDS	intrusion detection system
IEC	International Electrotechnical Commission
IEEE	Institute of Electrical and Electronics Engineers
IED	intelligent electronic device
IIMG	Interagency Incident Management Group
IM	instant messaging
IMANET	Internet-based mobile ad hoc networks
IN	inch
INFOCON	information operations condition

INHg	inches of mercury
INST	instantaneous
INSUL	insulation
INT	international
INTL	internal
IO	information operations or "indistinguishability obfuscation" (cryptographic term)
I/O	Input/Output
IoT	Internet of Things
IOPS	Input/Output Operations Per Second
IP	Internet Protocol
IPMASQ	IP masquerading
IPSec	Internet Protocol Security
IPS	iron pipe size
IPS	intrusion prevention system
IRP	Incident Response Plan
IRL	in real life
IS	information system
ISA	International Society of Automation
ISAC	Information Sharing Analysis Centers
ISCM	information security continuous monitoring
ISP	Internet Service Provider
ISSM	information system security manager
ISR	intelligence, surveillance, and reconnaissance
IT	information technology
IT-ISAC	Information Technology Information Sharing and Analysis Center
IV	indirect vent
IW	indirect waste
IWMS	Integrated Workplace Management System
J	joule
J-BASICS	Joint Base Architectures for Secure Industrial Control Systems
JBOD	just a bunch of disks
JEL	Joint Electronic Library
JIMS	joint incident management system
JMC	Joint Malware Catalog
JLLIS	joint lessons learned information system
JLLP	Joint Lessons Learned Program
JPAS	Joint Personnel Adjudication System
JT	joint test
K	kelvin
KBtu	one thousand British thermal units
KG	kilogram
KM	kilometer
KMP	key management personnel
KM²	square kilometer
KPA	kilopascal
KS	kitchen sink
KW	kilowatt
L	liter
LAN	local area network
LART	luser attitude readjustment tool

LAT	leaving air temperature
lb	pound
lbf	pound-force
LCM	luminaire control module
LCS	laboratory controls system
LE	law enforcement
LE/CI	law enforcement and counterintelligence
LEL	lower explosive limit
LES	law enforcement sensitive
LED	light emitting diode
Li-Fi	light fidelity
LIQ	liquid
LNS	LonWorks Network Services
LOTO	Lockout-Tagout
LP	low pressure
LRA	locked rotor amps
LSAP	link service access point
LSDU	link service data unit
LTD	long-time delay
LTEL	long-term exposure limit
LTO	linear tape open
LUN	logical unit number
LVL	level
LVPCB	low-voltage power circuit breaker
LVR	louver
LWT	leaving water temperature
M	meter
M^2	square meter
M&C	monitoring and control software
MAC	medium access control address, mission assurance category
MAN	manual
MAT	mixed air temperature
MAX	maximum
MB	megabyte
MBH	thousand British thermal units per hour
MCA	maximum overcurrent protection
MCCB	molded case circuit breaker
MD	motorized damper
MES	manufacturing execution system
MEVA	mission essential vulnerable areas
MERV	minimum efficiency reporting value
MFR	manufacturer
mg	milligram
mgd	millions gallons per day
MICR	magnetic ink character reader
MIN	minimum
MITM	man-in-the-middle attack
ml	milliliter
M	millimeter
M^3	cubic millimeter
MMS	Manufacturing Message Specification

MOA	memorandum of agreement
MODEM	modulator demodulator unit
MOU	memorandum of understanding
MPT	male pipe thread
M TYPE	lightest type of rigid copper pipe
MSBF	mean swaps between failure
MS/TP	master-slave/token passing
MTBF	mean time between failure
MT	mounted
MTTR	mean time to repair
MTU	master terminal unit
MU	make up
MV	machine vision
MVACB	medium-voltage air-magnetic circuit breaker
MVVCB	medium-voltage vacuum circuit Breaker
MW	megawatt

NA	not applicable
NAC	network access control
NAI	named area of interest
NAS	Network-Attached Storage
NAT	Network Address Translation
NC	normally closed, Numerical Control
NCIC	National Cybersecurity and Communications Integration Center
NCCG	National Cyber Response Coordination Group
NCSC	National Cyber Security Division
NDU	National Defense University
NE	negative
NERC	North American Electric Reliability Council
NESCOR	National Electric Sector Cybersecurity Organization Resource
NFC	near field communication
NFPA	National Fire Protection Association
NIAC	National Infrastructure Advisory Council
NIC	not in contract
NICCS	National Initiative for Cybersecurity Careers and Studies
NIDS	network intrusion detection system
NIPP	National Infrastructure Protection Plan
NIPRNET	Non-Secure Internet Protocol Router Network
NIR	network intelligence report
NISP	National Industrial Security Program
NISPOM	National Industrial Security Program Operating Manual
NIST	National Institute of Standards and Technology
NLZ	no-lone zone
Nmap	Network Mapper
NO	normally open
NOC	Network Operations Center
NO/NC	normally open/normally closed
NOSC	Network Operations Security Center
NPHP	name plate horsepower
NPPD	National Protection and Programs Directorate
NPS	nominal pipe size
NPSH	net positive suction head

NSC	Network Service Centers
NSP	Network Service Provider
NSTB	National SCADA Testbed
NTS	not to scale
NVD	National Vulnerability Database
O	oxygen
OA	outside air
OAT	outside temperature
OC	on center
OD	outside diameter
OE	open end duct
OEM	Original Equipment Manufacturer
OES	Office of Energy Infrastructure Security
OF	overflow
OI	operational impact
OEM	original equipment manufacturer
OLE	object linking and embedding, bullfighter call
OPC	object linking and embedding (OLE) for process control
OPORD	operation order
OPREP	operational report
OPSEC	operations security
OS	operating system
OSI	Open Systems Interconnection
OTP	one-time programmable bytes
OV	outlet velocity
OZ.	ounce
P2P	peer-to-peer
P&ID	piping and instrumentation diagram
PA	pascal
PC	plumbing contractor
PCF	point control function
PCHWP	primary chilled water pumps
PCI	Personal Computer Interconnect
PCL	personnel (security) clearance
PCR	pumped condensate return
PCT	programmable communicating thermostat
PD	pressure drop
PDU	power distribution unit
PDS	protected distribution system
PED	portable electronic device
PF	power factor
PFD	process flow diagram
PG	pressure gauge
PHC	preheat coil
PICS	Protocol Implementation Conformance Statement
PIDS	Protocol-based Intrusion Detection System
PII	personally identifiable information
PIN	personal identification number
PIT	platform information technology
PL	plate
PLC	Programmable Logic Controller

PNEU	pneumatic
POC	point of contact, proof of concept
PoE	Power over Ethernet
PP	protection profile
PPA	power purchase agreement
PPE	personal protective equipment
pph	pounds per hour
PRA	platform risk assessment
PRESS	pressure
PROP	propeller
PRV	pressure reducing valve
psi	pounds per square inch
psia	pound per square inch absolute
psig	pound per square inch gauge
psf	pounds per square foot
PSS	Plant Supervision Software
PUP	Potentially Unwanted Programs
PV	plug valve or photovoltaic
PVB	polyvinyl butyral resin
PVT	performance verification test
PWS	performance work statement
QAZ	a network worm
QTY	quantity
R2L	Remote-to-Local User
R^3	rapid recovery repairs
RA	response actions or return air
RAB	RAID Advisory Board
RAD	radius
RAID	redundant array of independent or inexpensive disks
RAM	random-access memory
RAT	remote access tool or return air temperature
RBO	responsible building official
RD	roof drain
RDP	redundant data path
R/E	return and exhaust
RECOV	recovery
RED	reducer
REF	reference
RFID	radio frequency identification
REQD	required
REV	revision
RF	radio frequency or return fan
RH	relative humidity
RIRN	remote incident response network
RISI	Repository of Industrial Security Incidents
RL	refrigerant liquid
RM	room
RMF	risk management framework
RoIP	Radio over IP
RPO	recovery point objective
RS	refrigerant suction

RSBAC	rule set based access control
RSN	Robust Security Network
RTN	return
RTOS	real-time operating system
RTU	remote terminal unit
RV	relief valve
S	switch
SA	supply air or situational awareness
SAISO	Senior Agency Information Security Officer
SAL	security assurance level
SAN	storage area network
SAPF	special access program facility
SAT	supply air temperature
SC	shading coefficient
SCADA	supervisory control and data acquisition
SCAP	Security Content Automation Protocol
SCH	schedule
SCHWP	secondary chilled water pumps
SCI	sensitive compartmented information
SCPT	Standard Configuration Property Type
SCSI	Small Computer System Interface
SD	smoke damper
SDT	saturated discharge temperature
SF	supply fan
SEC	secondary
SENS	sensible
SEO	search engine optimization
SEP	separate
SEQ	sequence
SER	series
SER	secure equipment room
SERV	service
SET	social-engineer toolkit
SF	service factor
SFA	security fault analysis
SFUG	Security Features Users Guide
SHDB	SHODAN Hacking Database
SHT	sheet
SI	international systems of units
SIEM	Security Information and Event Management
SIM	sensor interface module, subscriber identity module
SMS	Short Message Service
SMTP	simple mail transfer protocol
SNMP	Simple Network Management Protocol
SNVT	Standard Network Variable Type
SOC	Security Operations Center
SOD	separation of duties
SOL	solenoid
SoM	strength of mechanism
SOW	statement of work or scope of work
SP	static pressure, or NIST Special Publication

SPANs	SmartPhone Ad hoc Networks
SPC	statistical process control
SPOF	single point of failure
SPEC	specification
sq	square
sq ft	square feet
SQL	Structured Query Language
SRI	Solar Reflectance Index
SRTM	Security Requirements Traceability Matrix
SSA	Serial Storage Architecture
SS	stainless steel
SSH	static suction head
SSI	Synchronous Serial Interface
SSP	System Security Plan
SSR	solid-state relay
SST	saturated suction temperature
STD	standard, short-time delay
STH	static total head
ST&E	security test & evaluation
STEL	short-term exposure limit
STIG	Security Technical Implementation Guides
STL	steel
STR	secure telecom room
SUCT	suction
SPLY	supply
SV	service
SVH	static velocity head
SW	service weight
SWS	service water
TAB	HVAC testing, adjusting and balancing
TAN	transaction authentication number
TCP	Transmission Control Protocol
TCP/IP	Transmission Control Protocol/Internet Protocol
TCQ	tag command queuing
TD	temperature difference
TDH	total dynamic head
TEMP	temperature
TEV	thermostatic expansion valve
TFS	traffic-flow security
TFTP	Trivial File Transfer Protocol
TH	thermometer
THK	thick
TI	technical impact
TID	Threat Identification Database
TIR	telephone infrastructure rooms
TOU	time-of-use
TPC	two-person control
TPFDD	time-phased force deployment data
TPI	two-person integrity
TSP	total static pressure
TRO	tailored readiness option

TSP	total static pressure
TTP	tactics, techniques and procedures
TWR	airport control tower
U2R	user-to-root
UCPT	User-defined Configuration Property Type
UCS	Utility Control System
UDP	User Datagram Protocol
UESC	utility energy services contract
UF	under floor
UFAD	underfloor air distribution
UFC	unified facilities criteria
UH	unit heater
UMCS	utility monitoring and control system
UNVT	User-defined Network Variable Type
UPP	User Partnership Program
UPS	uninterruptible power supply
URL	Uniform Resource Locator
USB	universal serial bus
USRP	Universal Software Radio Peripheral
US-CERT	United States–Computer Emergency Readiness Team
V	volt
VAC	vacuum
VANETs	Vehicular Ad hoc Networks
VAR	value-added reseller
VAV	variable air volume
VB	vacuum breaker
VCI	vacuum cleaning inlet
VCL	vacuum cleaning line
VEL	velocity
VERT	vertical
VFD	variable frequency drives
VIB	vibration
VIPS	voice intrusion prevention system
VLAN	virtual local area network
VLP	virtual learning portal
VoIP	Voice over IP
VOL	volume
VPN	virtual private network
VSD	variable speed drive
VP	velocity pressure
VTR	vent through roof
W	watt
WAN	wide area network
WANET	Wireless Ad hoc Network
WAP	wireless access point
WARNORD	warning order
WB	wet bulb
WEP	Wired Equivalent Privacy
WiDi	Wireless Display
Wi-Fi	Wireless Fidelity

WiGig	Wireless Gigabit
WCO	wall cleanout
WG	water gauge
WH	water heater
WPA2	Wi-Fi Protected Access
WPAN	wireless personal network
WSN	wireless sensor network
WWW	World Wide Web
XIF	External Interface File
XML	Extensible Markup Language
XSS	cross-site scripting
ZNE	zero net energy

CHAPTER 28

Cyber Standards

Army Corps of Engineers Guide Specifications
13801   Utility Monitoring and Control Systems: since renumbered to UFGS 25 10 10
15951   Direct Digital Controls for HVAC and other Local Building Systems: since renumbered 23 09 23

Federal Information Processing Standard (FIPS) Publications
FIPS 140-2   Security Requirements for Cryptographic Modules
FIPS 181   Automated Password Generator (APG)
FIPS 185   Escrowed Encryption Standard
FIPS 196   Entity Authentication Using Public Key Cryptography
FIPS 197   Advanced Encryption Standard (AES)
FIPS 199   Standards for Security Categorization of Federal Information and Information Systems
FIPS 200   Minimum Security Requirements for Federal Information and Information Systems
FIPS 201   Federal Information Processing Standards

Committee on National Security Systems
CNSSI-4009   Committee on National Security Systems Glossary

Executive Order 12829: The National Industrial Security Program

Office of Management and Budget
OMB A-130   Management of Federal Information Resources

National Initiative for Cybersecurity Careers and Studies

National Institute of Standards and Technology (NIST) Special Publications
800-12   An Introduction to Computer Security: The NIST Handbook
800-18   Guide for Developing Security Plans for Federal Information Systems
800-27   Engineering Principles for Information Technology Security
             (A Baseline for Achieving Security)
800-30   Guide for Conducting Risk Assessments
800-32   Introduction to Public Key Technology and the Federal PKI Infrastructure
800-33   Underlying Technical Models for Information Technology Security
800-34   Contingency Planning Guide for Federal Information Systems
800-36   Guide to Selecting Information Technology Security Products
800-37   Guide for Applying the Risk Management Framework to Federal Information Systems:
             A Security Life Cycle Approach
800-39   Managing Information Security Risk: Organization, Mission, and Information System View
800-40   Guide to Enterprise Patch Management Technologies
800-41   Guidelines on Firewalls and Firewall Policy
800-47   Security Guide for Interconnecting Information Technology Systems
800-48   Guide to Securing Legacy IEEE 802.11 Wireless Networks

© Luis Ayala 2016
L. Ayala, *Cybersecurity Lexicon*, DOI 10.1007/978-1-4842-2068-9_28

800-51	Guide to Using Vulnerability Naming Schemes
800-53	Assessing Security and Privacy Controls in Federal Information Systems and Organizations: Building Effective Assessment Plans
800-57	Recommendation for Key Management
800-58	Security Considerations for Voice over IP Systems
800-59	Guideline for Identifying an Information System as a National Security System
800-60	Guide for Mapping Types of Information and Information Systems to Security Categories
800-61	Computer Security Incident Handling Guide
800-63	Electronic Authentication Guideline
800-66	An Introductory Resource Guide for Implementing the Health Insurance Portability and Accountability Act (HIPAA) Security Rule
800-69	Guidance for Securing Microsoft Windows XP Home Edition: A NIST Security Configuration Checklist
800-72	Guidelines on PDA Forensics
800-76	Biometric Specifications for Personal Identity Verification
800-82	Guide to Industrial Control Systems (ICS) Security
800-83	Guide to Malware Incident Prevention and Handling for Desktops and Laptops
800-84	Guide to Test, Training, and Exercise Programs for IT Plans and Capabilities
800-86	Guide to Integrating Forensic Techniques into Incident Response
800-88	Guidelines for Media Sanitization
800-90	Recommendation for Random Number Generation Using Deterministic Random Bit Generators
800-106	Randomized Hashing for Digital Signatures
800-111	Guide to Storage Encryption Technologies for End User Devices
800-113	Guide to SSL VPNs
800-114	User's Guide to Securing External Devices for Telework and Remote Access
800-115	Technical Guide to Information Security Testing and Assessment
800-121	Guide to Bluetooth Security
800-123	Guide to General Server Security
800-124	Guidelines for Managing the Security of Mobile Devices in the Enterprise
800-128	Guide for Security-Focused Configuration Management of Information Systems
NISTR 7628	Guidelines for Smart Grid Cyber Security

Network Working Group

RFC 3748	Extensible Authentication Protocol (EAP)

Get the eBook for only $5!

Why limit yourself?

Now you can take the weightless companion with you wherever you go and access your content on your PC, phone, tablet, or reader.

Since you've purchased this print book, we're happy to offer you the eBook in all 3 formats for just $5.

Convenient and fully searchable, the PDF version enables you to easily find and copy code—or perform examples by quickly toggling between instructions and applications. The MOBI format is ideal for your Kindle, while the ePUB can be utilized on a variety of mobile devices.

To learn more, go to www.apress.com/companion or contact support@apress.com.